Resurrecting the Granary of Rome

Ohio University Press
Series in Ecology and History
James L. A. Webb, Jr., Series Editor

Resurrecting the Granary of Rome

Environmental History and
French Colonial Expansion in North Africa

Diana K. Davis

OHIO UNIVERSITY PRESS

ATHENS

Ohio University Press, Athens, Ohio 45701
www.ohio.edu/oupress
© 2007 by Ohio University Press

Ohio University Press books are printed on acid-free paper ⊗ ™

20 19 18 17 16 15 14 13 12 11   5 4 3

Publication of this book was made possible in part by a University Cooperative
Society Subvention Grant awarded by the University of Texas at Austin.

Part of chapter 5 appeared in slightly different form as "Potential Forests: Degradation Narratives, Science, and Environmental Policy in Protectorate Morocco, 1912–1956," *Environmental
History* 10, no. 2 (2005): 211–38. Reprinted by courtesy of the Forest History Society.

Library of Congress Cataloging-in-Publication Data

Davis, Diana K.
 Resurrecting the granary of Rome : environmental history and French colonial expansion in
North Africa / Diana K. Davis.
    p. cm. — (Ohio University Press series in ecology and history)
 Includes bibliographical references and index.
 ISBN-13: 978-0-8214-1751-5 (cloth : alk. paper)
 ISBN-10: 0-8214-1751-7 (cloth : alk. paper)
 ISBN-13: 978-0-8214-1752-2 (pbk. : alk. paper)
 ISBN-10: 0-8214-1752-5 (pbk. : alk. paper)
 1. Human ecology—Africa, North—History. 2. Desertification—Africa, North. 3. Deforesta-
tion—Africa, North. 4. Africa, North—Environmental conditions. 5. Africa, North—His-
tory—1517–1882. 6. Africa, North—History—1882– 7. France—Colonies—Africa—History.
8. France—Colonies—Africa—Economic policy. I. Title.
 GF702.D38 2007
 304.20961—dc22

                                                                        2007007032

For J. E. H.

Always and Forever

This land, once the object of intensive cultivation, was neither deforested nor depopulated as today . . . . [I]t was the abundant granary of Rome.

J.-A.-N. Périer, *De l'hygiène en Algérie,* 1847

When the Arabs, the Muslim hordes, invaded North Africa . . . this land was laid to waste.

Henri Verne, *La France en Algérie,* 1869

If . . . we decide to fight until our climate is transformed [by reforestation], it will be wealth, it will be life, it will be Algeria returned to its original fertility: it will be Algeria becoming the granary of France!

Bulletin de la Ligue du Reboisement de l'Algérie, 1882

Desertification . . . is uniquely the act of humans . . . . [T]he nomad has created what we call the pseudo-desert zone.

Louis Lavauden, "Les Forêts du Sahara," 1927

The Roman Empire's bread-basket in North Africa, which once contained 600 cities, is now a desert.

UN Chronicle, 1997

Contents

Illustrations

Maps

Color Plates

Following page 88

Preface and Acknowledgments

This book explores the environmental history of the Maghreb and argues that it is really only a story—a story first told early in the period of French occupation to facilitate colonial rule. The need for this work became apparent to me during my initial period of research in Morocco from 1995 to 1996, and again during a second round from 1997 to 1998. I was studying nomads at the edge of the Sahara; an analysis of their indigenous knowledge of veterinary medicine and of the environment formed the primary subject of my dissertation in geography on the political ecology of pastoralism. Fresh from seminars in arid lands ecology, geography, and pastoral societies, I arrived in the field with only a basic knowledge of French colonial history. It soon became clear to me, however, that many of the contemporary environment and development projects put in place by the Moroccan government, international agencies, and nongovernmental organizations (NGOs) were based, in large part, on colonial examples and colonial-era environmental data. All too often it was the nomads and other pastoralists who were blamed for environmental degradation dating back several centuries.

Despite the dire claims of overgrazing, deforestation, desertification, and environmental catastrophe repeated by the Moroccan government and several international institutions, the nomads in southern Morocco reported that their environment had not much changed during the last several decades and that it was neither degraded nor desertified. They further explained, with sophisticated reasoning, that changes in the environment (especially the vegetation) depended mostly on rainfall, and that what might appear as degraded bare ground would become a carpet of healthy plants with the next good rain. The scant contemporary data available tended to support the environmental story of the nomads rather than that of the government or the NGOs. In the course of completing my dissertation, I frequently found versions of what I describe in this book as the French colonial declensionist environmental narrative, that is, a tale of environmental decline since what was presumed to have been the fertile and forested Roman period. This colonial narrative primarily blamed the pastoralists, especially nomadic pastoralists, and their "ancestors," the "invading Arabs" of the eleventh century, for deforesting and desertifying North Africa. My work on contemporary Morocco thus opened the door to the present project, a deep history of the Maghreb and its environment under French colonial administration.

This study examines how the colonial declensionist environmental narrative was created and used during the period of French rule in Algeria, Tunisia, and Morocco, from the French conquest of Algeria (1830) to the independence period of all three colonies. The main argument is that the declensionist narrative replaced an earlier narrative of lapsed fertility in order to justify and enable the appropriation of land and resources, social control of local populations, and transformation of subsistence production into market-integrated production. Late in the colonial period this narrative became enshrined in ecological science—and it is this legacy that is the most important and long-lasting, since it continues to inform environmental research and policy in the Maghreb and much of the broader Mediterranean basin to the present day.

I am in this sense presenting, among other things, one argument for the continued significance of the Maghreb, a region that has not received the attention it merits in recent studies in environmental history, geography, and related disciplines. Although I do so for the most part in the context of an environmental history, I have taken an interdisciplinary approach incorporating the literatures and methods of several fields, including geography (specifically political ecology), North African history, and French colonial studies.

This book is based primarily on one and a half years of (nonconsecutive) research in the French colonial archives. I am very grateful for research funding from the National Endowment for the Humanities (NEH) Summer Stipend Program and two Summer Research Awards from the College of Liberal Arts, the University of Texas at Austin, which made research and travel for this book possible. Time for much of the writing was generously provided by a Dean's Fellowship from the College of Liberal Arts, the University of Texas at Austin. A University Cooperative Society Subvention Grant awarded by the University of Texas at Austin helped to defray production expenses and, importantly, made the many illustrations possible.

For his enthusiasm for the project since its early stages I am deeply grateful to James L. A. Webb, Jr., the editor of the Ohio University Press Series in Ecology and History. Senior Ohio editor Gillian Berchowitz has been a joy to work with, and I thank her warmly for all her help and expertise. The production team at Ohio, especially Rick Huard and Beth Pratt, similarly deserve my thanks for their expert assistance.

I have benefited enormously over the years from many people who helped me in various ways with this project. I owe a great intellectual debt, as well as many thanks, to Iain Boal, Terry Burke, Louise Fortmann, David Hooson, Lynn Huntsinger, Ian Manners, and Michael Watts. For stimulating discussions, suggestions, and other assistance that helped to shape different parts of this project, I thank Peter Abboud, Ellen Amster, Amman Attieh, Stephen Bell, Roger Byrne, Shana Cohen, Lisa Corti, Denis Cosgrove, Kate Davis, Paul English, Edouard le Floc'h, Denys Frappier, Mia Fuller, Stephanie Goodman, Derek Gregory, Jennifer Grocer, Katherine Hoffman, Robert Holz, Henry Le Houérou, Lisa Husmann, Doug Johnson, Susan Jones, Mustapha Kamal, Annelisa Kilbourn, Maria Lane, Elizabeth Lawrence, Franklin Loew, Margaret Mackenzie, Kathy McAfee, James McCann, James McCarthy, Larry Michalak, Jim Miller, Don Mitchell, Liz Oglesby, Stefania Pandolfo, Ying Yang Peterson, Ravi Rajan, Frances Robertson, Emery Roe, Adam Rome, Amy Ross, Jim Ross, Sharmila Rudrappa, Nezar Al-Sayyad, Chuck Schmitz, David Sherman, Susan Slyomovics, Neil Smith, Al Sollod, Paul Starrs, Chip Stem, Will Swearingen, Liz Vasile, Barbara Walker, Peter Walker, William Wier, and Wendy Wolford. I would especially like to thank Paul Claval for his suggestions and James Housefield for his expert assistance in analyzing much of the art discussed in this book.

For their long-standing friendship and support, I am grateful to Karl Butzer, Shane Davies, Robin Doughty, Greg Knapp, Ian Manners, and Francisco Perez, all in the geography department at the University of Texas at Austin. Bill Doolittle, particularly, deserves my thanks for the many ways he helped with this project. The rest of my fellow faculty members have likewise been supportive of this project, for which I thank them. New colleagues at UT have given generously of their time, shared ideas, and helped in many other ways, particularly Kamran Aghaie, Kamran Ali, Janet Davis, Charlie Hale, Hillary Hutchinson, Brian King, Mark Lawrence, Roger Louis, Abe Marcus, Esther Raizen, Dina Sherzer, Keith Walters, and Leo Zonn. Dee Dee Barton and Nicole Morales deserve special thanks for their assistance and good humor in helping me to navigate many of the administrative mazes related to this project. Others at UT who have been instrumental in helping to bring this book to fruition include Connie Brownson, Sheldon Eckland-Olson, Richard Lariviere, Juan Sanchez, and Joey Walker. The library resources and staff at UT were extremely helpful. I am grateful to Nancy Elder, Susan Macicak, and Paul Rascoe for their help. The wonderful staff at interlibrary loan services, expertly directed by Wendy Nesmith, deserves my thanks for their unflagging and expert help obtaining several obscure books and articles. In particular, I would like to thank David Reed, Brad Sunday, and Donna Terpack-Palter. For their energetic assistance during the final months of writing, I owe a big "thank you" to Heather Allard, Claire Altman, Caroline Carter, Bailey Hayes, Wassia Khaja, and Ebony Porter.

I am indebted to many who helped me, in France, at the Bibliothèque Nationale de France (BNF), Paris; Centre des Archives Diplomatiques de Nantes (CADN), Centre des Archives d'Outre-Mer (CAOM), Aix-en-Provence; Centre d'Écologie Évolutive et Fonctionelle (CEFE), Montpellier; Centre des Haute Études sur l'Afrique et l'Asie Modernes (CHEAM), Paris; Centre de Coopération Internationale en Recherche Agronomique pour le Développement (CIRAD), Montpellier; the library at the École Nationale du Génie Rural des Eaux et des Forêts (ENGREF), Nancy; the Institut du Monde Arabe (IMA), Paris; Institut de Recherches et d'Études sur le Monde Arabe et Musulman (IREMAM), Aix-en-Provence; the library at the Muséum National d'Histoire Naturelle, Paris; and the Service Historique de l'Armée de la Terre (SHAT), Paris. For their expert and friendly assistance, I especially need to thank Mme. Malecot at CHEAM, Marie-Jeanne Lionnet and David Gasparotto at ENGREF, Damien Heurtebise at CADN, and Catherine Régnault at the Musées des Beaux-Arts de Rouen. I would also like to thank the Association Paul-Albert Février in Aix-en-Provence for material assistance with this project. Outside France, I am likewise grateful to those who helped me at the India Office of the British Library, London; Center for Southwest Research, University of New Mexico, Albuquerque; Humanities Research Center at the University of Texas at Austin; and the Hunt Institute for Botanical Documentation, Carnegie Mellon University, Pittsburgh. Angela Todd at the Hunt Institute deserves my special thanks. I thank the Forest History Society for allowing me to include a revised version of my article, "Potential Forests: Degradation Narratives, Science, and Environmental Policy in Protectorate Morocco, 1912–1956," as part of chapter 5.

As with many endeavors in life, the support of family was crucial at several points in the research and writing of this book. First and foremost, I must thank the living beings who have given me the honor of sharing their lives with me: James Housefield, Max Davis-Housefield, Manon, Thelonious, and Chewy. Thanks for the travels and adventures (and patience) that went into researching and writing this book, and for the support and encouragement all of you have given me over the years. I would like to thank David, Jan, Melissa, and Michael Davis for their patience, encouragement, and help. My late grandmother, Basima Davis, although frequently worried about my numerous and lengthy sojourns in faraway places, was always supportive of my quest for knowledge, for which I am eternally grateful. My adopted family also deserves my thanks, particularly Betty and Ken Housefield, Jean Ann and John Schingel, and Jeff and Jennifer Tidwell. The bad jokes and technical expertise of John and Erin Housefield (respectively) were especially important in keeping me going over the last few years.

All errors of fact or interpretation are, of course, my own.

CHAPTER 1

Imperial Stories and Empirical Evidence

THE ENVIRONMENTAL HISTORY of North Africa is a sad tale of deforestation and desertification that has spanned much of the past two millennia. This history of environmental decline has been recounted so often by so many that it is widely accepted without question today. Yet recent paleoecological evidence and new research in arid lands ecology do not support many of these claims regarding deforestation, overgrazing, and desertification. A closer examination of how the environmental history of North Africa has been constructed over time reveals the key roles of French colonial scientists, administrators, military men, and settlers in writing this declensionist narrative. The complex, dynamic and long-standing relationship between French colonialism, environmental narratives, and history in North Africa forms the primary subject of this book.

Scholars of colonialism have effectively documented the multiple ways in which the French administration expropriated land, forests, and other natural resources from North Africans during the colonial period.[1] What has been less well explored, however, is how the French environmental history of North Africa, and environmental and related laws and policies, were used to facilitate the appropriation of these resources, to transform subsistence production, and to effect social control. By detailing the construction and use of the declensionist environmental narrative, this book tells the story of the French colonial story of nature in the Maghreb.[2]

The conventional environmental history of North Africa most widely accepted today was created during the French colonial period. Before the conquest of Algeria, North Africa had been most commonly depicted in French and European writings as a fertile land that had lapsed into decadence under the "primitive" techniques of the "lazy natives." This view changed under French rule of the Maghreb, which began in 1830 with the occupation of Algeria.[3] In less than two decades, there emerged a colonial environmental narrative that blamed the indigenous peoples, especially herders, for deforesting and degrading what was once the apparently highly fertile "granary of Rome" in North Africa. The declensionist story that quickly developed was used throughout the colonial period to rationalize and to motivate French colonization across North Africa. This narrative and its utilization reached their apogee between 1880 and 1930, precisely the period during which colonial activities caused the most deforestation. Although the colonial narrative began to take shape first in Algeria, it included the entire Maghreb environment from its inception, and it was quickly applied to the subsequently conquered territories of Tunisia (1881) and Morocco (1912). This story was partly a political history and partly an elaborate environmental history of the preceding two millennia. Classical literary sources, including the writings of Herodotus, Pliny, Procopius, Strabo, and Ptolemy, had formed the basis for French and European views of North Africa as "the most fertile region in the world" long before the French conquest of Algeria.[4] By the second half of the nineteenth century, the story became firmly entrenched that North Africa had been the granary of Rome. This exaggerated image was reinforced by the archeological examination of many Roman ruins, including aqueducts and other irrigation structures as well as ruined towns and olive presses, which appeared to have supported large and prosperous populations. Orientalist paintings like Victor-Pierre Huguet's 1868 *The Remains of a Roman Aqueduct in the Region of Cherchell* often represented these traces of the granary of Rome in ways that fed the colonial imagination (see plate 1).

Later in the colonial period, the image of the successful exploitation of the natural fertility of North Africa by the Romans was accompanied by an image of the subsequent destruction, deforestation, and desertification of the North African environment by hordes of Arab nomads and their ravenous herds. Hints of this transformation in the narrative are evident as early as the 1830s. The major transformation to a declensionist narrative, though, did not appear widely until the 1850s and 1860s. By the 1870s, it was ubiquitous. Many official and popular writers on the Maghreb essentially agreed with the conclusion of the Tunisian Service des Affaires Indigènes, that "the profound convulsions which, since the Roman era, have upset the country: the passage of the Arab armies and later the Hillalian tribal invasion . . . have made of this country a desert strewn with ruins which, however, attest to its ancient prosperity."[5]

While the French largely drew on classical sources for evidence of North Africa's former fertility and vast forests, they relied on the writings of medieval Arab historians to support their tale of decline at the hands of Arab nomads and their herds over the prior eight centuries. Small portions of Ibn Khaldoun's voluminous writings, for example, selectively chosen for their negative view of Arab nomads, became frequently cited sources for many of the French colonial claims of ruin wrought by the Hillalian Arab nomad "invasion" of the eleventh century. Although other Arab writers had been cited often in the first two decades of occupation, after the French translation of Ibn Khaldoun's *Prolégomènes* and *Histoire des berbères* by Baron de Slane in the 1850s most French writing on North Africa quoted Ibn Khaldoun in its descriptions of the widespread destruction caused by the Arab invasion. Many such citations spoke to the destruction of "civilization," that is, of urban areas and urban-based political and economic systems.

Authors writing in particular about the devastating effects of the Arab invasion on "civilization" did not, or did not often, mention its environmental effects. It is not generally recognized, though, that a few select passages from Ibn Khaldoun on the environmental destruction wreaked by the Hillalian invasion were in fact identified, widely quoted, and in the process reified by those invoking the declensionist narrative. Thus, for the last half of the nineteenth century and the first several decades of the twentieth, Ibn Khaldoun was cited repeatedly for his description of the Arab nomads in the eleventh century as "locusts" who "ruined gardens and cut down all the trees."[6] Even widespread forest fires, and thus presumed deforestation, were attributed to the Hillalian nomads based on citations of Ibn Khaldoun.[7] Ibn Khaldoun was also quoted frequently for irrefutable proof that "the Arab nomads brought devastation" and that "civilization was ruined and the country was changed into a desert."[8]

Fears of desertification were extremely widespread throughout the colonial period in the Maghreb. As early as 1834 military officers in Algeria were blaming nomads for destruction of the vegetation and of the soil itself.[9] Many French colonists were describing Algeria by the 1860s as the "land of thirst," a land that had been transformed from a former paradise into a sterile, barren desert. The word "desertification," though, does not appear to have been used until 1927, when Louis Lavauden described the forests of the Sahara as "desertified" and discussed "desertification" as a uniquely human act.[10] As early as 1880, however, direct and detailed blame was being placed on North Africans, especially nomadic pastoralists, for what was later called "desertification." A quarter of a century later, in 1906, one of the foremost French experts on North Africa exclaimed that, due to burning and overgrazing in Algeria, "the forest gives way to scrub, the scrub to herbaceous vegetation, the herbaceous vegetation to bare soil, that finishes by being detached itself and that becomes the victim of the wind."[11] This description of desertification became increasingly common during the rest of the colonial period and still informs writing on desertification today. Such descriptions often preceded pronouncements about the former fertility of the Roman period and the need for France to resurrect it. One typical narrative, for instance, asserted that "following the prosperous days of the Roman era, Algeria, Tunisia, and Morocco . . . vegetated, barely surviving. . . . All fell into ruin, of the splendid golden age of Roman Mauretania only a charred rocky desert remained. . . . Suddenly, a new breeze, and France . . . arrived to conquer so that Latin civilization could be rescued."[12]

The French colonial conception of Roman North Africa as spectacularly fertile and prosperous rested, in large part, on the belief that the Maghreb had been the granary of Rome and that the many ruins indicated a large population thriving in antiquity. Since the Roman ruins were no longer being used when the French colonized the region, and the local population was relatively small, they assumed that some sort of environmental catastrophe had occurred, and primarily blamed the Arab nomads for the perceived environmental degradation. The idea that the Maghreb had supplied grain to the Roman Empire is well supported by the historical record. The French colonial belief that the Maghreb had produced significantly more grain during the Roman period than afterward, however, is not well supported by the available evidence.

It is true that significant quantities of grain were produced in and exported from North Africa from before the Roman period until the twentieth century.[13] During the Roman period it is estimated that five million bushels of grain were shipped annually from North Africa to Rome.[14] By comparison, however, in 1862, Algeria alone produced 34.5 million bushels of grain, 90 percent of which was grown by Algerians using, significantly, traditional techniques.[15] By 1954, Algeria produced around 86 million bushels of grain, nearly half of which was sold on the international market.[16] As populations have grown in the Maghreb and other regions of the Mediterranean have begun to produce more grain, exports of cereals have declined, but production has continued to rise.[17]

A significant amount of land degradation did, however, occur during the Roman period itself, as a result of their agricultural techniques and expansion. Contemporary research in fact attributes the blame for the commencement of soil degradation in North Africa to the Romans, whose overcultivation was "followed [by] a phase of relative soil conservation and vegetative regeneration with the more nomadic land-use system of the Arabs."[18] Furthermore, these authors conclude that "conservation of Mediterranean landscapes . . . can only be ensured by continuation of the agro-pastoral functions under which these landscapes evolved."[19]

The deforestation and desertification assumed to have been wrought by the Maghreb's large pastoral population, especially since the eleventh century, were alleged to have continued unabated into the nineteenth century. Therefore, livestock raisers, especially nomads, received the vast majority of the blame for what the French interpreted as centuries of environmental abuse of a previously lush landscape. From the story of North Africa's having been the bountiful granary of Rome and the tale of subsequent ruin and deforestation due to the Arab nomad "invasion," the French fashioned a justification and an imperative for their colonial projects.[20] They told themselves that, to restore the former glory and agricultural fertility of Rome, they must save North Africa from the "destructive natives."[21] Jean Colin, for example, taught the young recruits of the Indigenous Affairs Service that "France is the legitimate successor of Rome. . . . The great Roman people of whom we are the heirs conquered this region well before the Arabs."[22] He later advised the recruits that, "like Rome, we will again expand the cultivable area . . . and transform [it] into fertile plains."[23] The cultivable area was indeed expanded across the Maghreb, as was the afforested area. One influential colonist exhorted his countrymen to plant trees everywhere by promising that "if . . . we decide to fight until our climate is transformed [by reforestation], it will be wealth, it will be life, it will be Algeria returned to its original fertility: it will be Algeria becoming the granary of France!"[24]

The need for reforestation of areas assumed to have been deforested by the "natives" was commonly used as an excuse to appropriate land. Armed with the colonial environmental narrative, the French passed new laws and policies as early as the 1830s to curtail and criminalize many of the traditional uses of the environment by the Algerians. Colonial laws on property and land tenure, as well as laws on forestry and grazing, not only transformed the use of the land but also effected the appropriation of large amounts of land and resources for the settlers and the French administration. Many of the most important and far-reaching of these laws and policies, as demonstrated in the following chapters, were justified with the declensionist narrative. The narrative was so ubiquitous and influential, in fact, that it was actually written into several of these laws, including the Algerian Forest Code.

Figure 1.1. Roman ruins of Cuicul at Djemila, Algeria. The original caption of this photo states, "Methodical excavations begun in 1909 have revealed at Djemila, on the location of the ancient Roman town Cuicul, in a *hilled site today desert* [*désertique*], a group of very interesting Roman monuments." This caption insinuates that the region was more humid and vegetated during the Roman era, thus implying an environmental decline over the last two millennia. From Clément Alzonne, *L'Algérie* (Paris: Fernand Nathan, 1937), 25. Reproduced with the kind permission of Armand Colin.

Figure 1.2. The High Plateaus near Constantine, Algeria. Reinforcing the colonial environmental narrative that blamed Arab herders for land degradation, the original caption of this photo reads, "High Plateaus—region of Batna. Forests in regression, surrendered to grazing." From Henri Marc, *Notes sur les forêts de l'Algérie*, Collection du centenaire de l'Algérie, 1830–1930 (Paris: Larose, 1930), planche 14. Reproduced with the kind permission of Maisonneuve & Larose.

Images such as the photograph shown in figure 1.1 frequently illustrated popular and official writings on North Africa. Hinting at the declensionist narrative, the caption of this photo from a popular book reads, "Methodical excavations begun in 1909 have revealed at Djemila, on the location of the ancient Roman town Cuicul, in a *hilled site today desert [désertique]*, a group of very interesting Roman monuments."[25] Other images, especially in forestry publications, were more explicit in assigning blame for assumed environmental destruction. Figure 1.2 shows a photograph from an official government forestry book of the region of the High Plateaus near Batna. The caption reads, "Forests in regression, surrendered to grazing."[26] In this case it is quite clear that the livestock of local herders were held responsible for deforestation. Images and descriptions thus worked together throughout the colonial period to reinforce the dominant environmental narrative, of the deforestation and desertification of a formerly fertile and wooded land.

The claim of the previous existence of large and lush forests was an important part of the story of North Africa's ancient fertility and more humid climate. Early descriptions of forests came mostly from Greek and Roman sources. In 1846, for example, one French author quotes Strabo's assertion that "all of the [land] situated between Carthage and the Pillars of Hercules (from Tunis to the [Atlantic] ocean) is of an extreme fertility" and then quotes Pliny on "the grand forests with which the sides [of the Atlas Mountains] are covered."[27] This practice of selective quotation was common well into the twentieth century. Just prior to the conquest of Morocco, similar descriptions were found frequently in French scholarship on the territory. "The Atlas were covered with thick forests," and the trees were "amazing . . . gigantic," wrote one of the members of the scientific mission to Morocco, citing Pliny.[28] Large forests were believed to have once covered not only North Africa's mountains and coastal areas but also many desert regions. French foresters working in Algeria noted that, even in arid regions of the High Plateaus and the Saharan Atlas, as far south as Laghouat, there were "numerous traces of ancient forests."[29]

The importance of forests in the dominant environmental narrative grew over time—as did estimates of deforestation. Although the extent and composition of forests in Algeria were not known with any certainty until the early twentieth century (and in Morocco not until the 1930s), estimates of deforestation and the consequent pressure for reforestation increased during the late nineteenth and early twentieth centuries. Aided by the young science of phytosociology (plant ecology), these exaggerated estimates of deforestation became institutionalized as scientific fact in the 1920s and 1930s. By the end of the colonial period, it was widely accepted in the Maghreb, in France, and in much of the world that North Africa had suffered a loss of between 50 and 85 percent of its original forests over the preceding two thousand years.[30]

In many scientific, administrative, and policy circles similar statistics are invoked today. Deforestation and desertification are still commonly believed to have occurred for millennia in North Africa and throughout much of the Mediterranean basin. The long historical scientific view available now, however, has shed some revealing light on the environmental history of the Maghreb and the surrounding regions.[31] It has been established, for example, that during the last Ice Age (Pleistocene) many regions of the Mediterranean and Middle East were not forested, due to dry and cold climatic conditions. Rather, large areas of grassland and steppe predominated.[32] In fact, during the last Ice Age the forested area of the globe was only about 32 percent of what it is today, and the total unforested area of the world was at least 1.5 times as large as it is today.[33] These sources show that in the Mediterranean basin a general pattern of expanding forests did not occur until approximately the ninth to the seventh centuries BCE. This was followed by alternating periods of relatively more arid and relatively more humid conditions, which varied considerably from region to region, until approximately 1000 BCE. It is widely agreed that climatic and vegetation conditions then stabilized in the more arid pattern still found today.[34] Thus, the available physical evidence provides a view of the environmental history of the Mediterranean basin and surrounding regions very different from that developed by the French in the nineteenth century. It appears that there has been a long history of a comparatively treeless landscape with a dynamic and migrating vegetation.

Physical evidence specific to the Maghreb points to a more humid climate that settled in around 4000 to 3000 BCE and lasted with a few variations until approximately 1000 BCE, when a much more arid and stochastic climate (similar to that of today) became the norm.[35] This humid period produced a much wetter environment in what is now known as the pre-Saharan regions, supporting such tropical and subtropical animals as elephants and rhinoceros.[36] Evidence from fossil pollen samples taken in northeastern Algeria shows signs of this wetter phase: forest vegetation dominated the local landscape until approximately 2000 BCE, when it began to diminish and steppe vegetation and grasses began to increase.[37] In the majority of these North African pollen cores, the levels of grass pollen have remained remarkably constant, with some fluctuations, from the Pleistocene until the present, indicating that widespread desertification has not occurred.[38] The existing evidence of climate change and vegetation history in the region have led many to the conclusion that it is difficult to distinguish between man-made degradation and a natural trend toward aridification over the last two to three thousand years, especially at lower elevations.[39]

Morocco has the highest rate of forest cover in the Maghreb and is usually considered the most severely deforested, historically, of the three countries. Morocco also has had more paleoecological research conducted within its borders than the other Maghreb countries. The pollen core data published to date show that since the appearance of trees in the mountains of Morocco between three thousand and eighty-five hundred years ago there have been fluctuations in tree cover, including some deforestation in a few areas but not in others. In the Middle Atlas region, four separate pollen cores, all of them carbon 14–dated, show some fluctuations in the pollen of oaks, cedar, pine, and other trees over the last two thousand years, but reveal no significant trend in any direction.[40] In the case of pine, a large decrease is shown in two of these cores, but they occur at about two thousand to seventeen hundred and fifty years ago, long before either of the "Arab invasions." Likewise, deciduous oak appears to have declined steeply in two of the cores, but this occurred around three thousand years ago and likely coincided with the drier climate that settled in at that time. Interestingly, some of these cores also show increases in species such as juniper, pine, and cedar over the last several hundred years. Evergreen oak is remarkably stable. A similar pollen core is available from the High Atlas Mountains, also carbon-dated.[41] This core shows fluctuations of pine, oak, and cypress but no significant change over the last three thousand years. Overall, then, the samples that have been accurately carbon-dated show no definitive overall pattern of massive deforestation on the order of the frequently claimed 50 to 85 percent over the last two millennia.[42]

In Algeria and Tunisia the pollen core data are more limited, but the studies that exist suggest that decreasing trends in the pollen of a few species like pine and some oaks began about four thousand years ago, well before the Roman period.[43] Many other species, though, have been fluctuating without trend, and some, such as cypress and certain oaks, particularly cork oak, have even been increasing in several of these areas over the last two thousand years.[44] This is very similar to the evidence for forestation and deforestation in the wider Mediterranean basin. As one scholar has explained, after the appearance of trees in the mid-Holocene "most [pollen] diagrams show either stasis, or fluctuations with no indication of trend, in arboreal pollen."[45] Even data from the Algerian mountains of the central Sahara show no significant change in vegetation over the last four thousand years.[46] Moreover, these recent data indicate that old, dead tree stumps are not necessarily "traces of ancient forests," as many colonial foresters believed.[47] What nearly all of the Maghreb data shows is that the significant changes in tree pollen took place thousands of years before present, and well before either of the "Arab invasions." On the basis of her pollen analyses in Tunisia, Annik Brun has challenged the "Arab invasion" hypothesis of deforestation so popular with some French authors.[48] She proposes instead that it was more likely that general aridification has favored a vegetation that supported an expansion of nomadism over the last two thousand years. Much of the paleoecological data for the Maghreb in fact shows a general trend of aridification over the last fifteen hundred years, making it difficult, if not impossible, to conclude that vegetation changes regarded as degradation are solely the result of human ac-

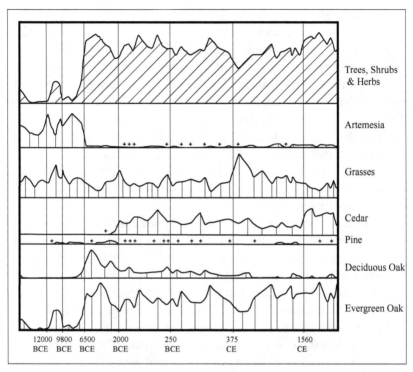

Figure 1.3. Diagram of pollen core data from Lake Tigalmamine in the Middle Atlas Mountains, Morocco. This diagram illustrates changing levels of plant pollen over the last 14,000 years. It shows a clear and significant increase in tree pollen about 8,500 years ago. Since that time, several shifts in tree pollen have occurred, but no significant trend toward either increased or decreased pollen levels is evident, indicating that no major changes in tree populations are likely to have occurred. The most significant decrease in pollen appears to have taken place about 1,600 years ago, well before either of the "Arab invasions." Details of cedar, grass, and evergreen oak pollen are also illustrated here. Based on H. F. Lamb, U. Eichner, and V. R. Switsur, "An 18,000-Year Record of Vegetation, Lake-Level and Climate Change from Tigalmamine, Middle Atlas, Morocco," *Journal of Biogeography* 1, no. 18 (1989): 71. Diagram by D. K. Davis.

tivity until the colonial period.[49] Indeed, in the words of Brent Shaw, "by far the greatest proportion of the loss of North Africa's forests took place within the last century, primarily the half-century between 1890 and 1940," well after the French had formed their declensionist story of the environmental history of North Africa.[50]

This is not to say that no deforestation took place in the Maghreb before the colonial period. On the contrary, trees have certainly been felled for human use and destroyed by natural forces such as fire and disease over thousands of years. It is vitally important, however, to place what deforestation likely did occur in an appropriate context. None of the colonial estimates, for instance, took natural regrowth of arboreal vegetation into account when making evaluations of North African forests. Instead, estimates of deforestation implied a somehow permanent loss of the trees believed necessary for civilization. As is detailed in the appendix, though, much of the vegetation in the Maghreb is highly resilient and regrows vigorously after disturbances, including fire, grazing, or cutting. The paleoecological research conducted in the region to date supports the historical regrowth of many of the trees and other vegetation that may have been removed at one time or another.

Paleoecological research such as fossil pollen analysis, however, was extremely limited during the colonial period, appearing only in the mid-twentieth century.[51] From the 1830s through the early twentieth century the dominant method of reconstructing past vegetation was to compare historical literary descriptions to contemporary conditions visible in the landscape. Small clumps of trees, for example, were assumed quite often to be the remnants of large forests that had been destroyed. Such "relict vegetation" was then used to deduce what the "natural" vegetation should and could be in a region. Based on comparisons with written sources, especially with those of the Roman period described above, the conclusion was drawn by most that the Maghreb as the French found it was egregiously deforested.[52] Such comparisons date from the first years of colonial occupation in Algeria and were made throughout the colonial period in each of the three Maghreb colonies. Moreover, they formed the basis of most phytoecological research on North Africa in the early twentieth century. Yet both the concept of relict vegetation as applied here and a reliance on written historical sources to deduce natural vegetation are fraught with serious problems. Knowledge generated with both methods has been shown frequently to contain large biases, and even serious errors, based on the subjective judgments of individuals and groups.[53] In North Africa, the colonial environmental narrative strongly biased research on the state of the environment using these methods well into the twentieth century.

Despite the fact that this declensionist narrative became the dominant environmental history over the course of the colonial period, there were some voices of dissent in the nineteenth and twentieth centuries. This dominant narrative was primarily championed by those involved in the North African colonies in some way: administrators, foresters, colonists, military officers, businessmen, and the colonial lobby in France. It was sometimes partially challenged by members of the anticolonialist lobby, botanists, and a few French writers such as Stéphane Gsell. Gsell, of the Collège de France, thought it probable that "in antiquity as in our time, there were in Barbary [North Africa] vast regions denuded [of vegetation]" but that "there were also in this region extensive forests."[54] Gsell did subscribe to other parts of the dominant narrative, however, such as the substantial deforestation allegedly caused in some areas by the Hillalian nomad invasion of the eleventh century.[55] There were also occasional officers in the indigenous affairs bureaus and Saharan experts who had some respect for the indigenous knowledge and traditional practices of the local Algerian farmers, herders, and nomads. When it came to making policy and enacting and enforcing legislation, however, their voices were generally not heeded, and the dominant colonial environmental narrative remained one of decline and destruction.

The story did, however, change over the course of the colonial period, and the purposes to which it was put also varied depending on who was using it and why. What began as a lamentation over untapped fertility quickly became a declensionist narrative that blamed the Algerians and their ancestors for environmental destruction in order to further the colonial project.[56] Although it is difficult to identify exactly who wrote which part of this complex environmental history, it is possible to identify many of those who used it, and for what reasons, over time. The remainder of this book teases out these details in the context of the history of the complex relationship between environmental narratives and French colonialism in North Africa. The declensionist narrative became the standard environmental history by the end of the colonial period, and it remains dominant today. It helped to delegitimize traditional ways of living with the land in North Africa and to facilitate colonial expansion, and in the process to dispossess North Africans from the best of their lands and from most aspects of their traditional livelihoods. As a result, generations of North Africans were dislocated economically, politically, culturally, and physically. To comprehend just how profound these dislocations were, it is vital to understand a few basic facts regarding the environment and ecology of the Maghreb.[57]

Due to a combination of factors, including topography, latitude, altitude, and its relationship with the Atlantic Ocean, the Mediterranean Sea, and the Sahara Desert, the vast majority of the Maghreb is arid or semiarid. Most of southern Algeria is hyperarid, and small pockets along the

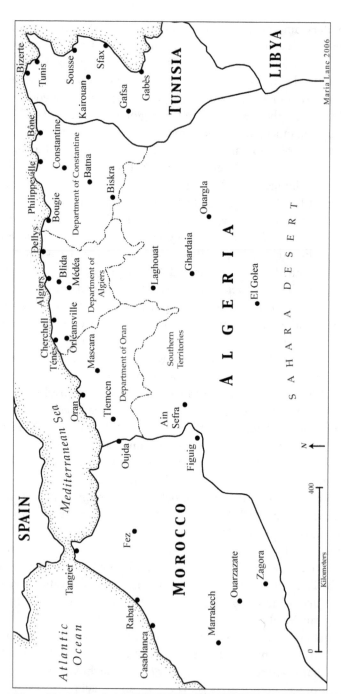

Map 1. The Maghreb. Modified after multiple sources.

coast or in high mountains are subhumid. Approximately 75 percent of the Maghreb receives 350 mm of rainfall or less annually. Summers are generally hot and dry, and the rain, which falls mostly during the winter months, varies a great deal from year to year. Droughts and wildfires are common.

The vegetation, having evolved over thousands of years along with livestock under these conditions, is largely well-adapted to grazing, as well as to drought, aridity, and fire. The ecology of the region is by and large resilient, and such traditional livelihood activities as the use of small fires for agricultural purposes were ecologically appropriate. Furthermore, traditional methods of raising livestock—including movement over large areas during much of the year, or pastoral nomadism—are now understood to be some of the most appropriate and sustainable uses to which these arid and unpredictable environments can be put.[58] Indigenous herders generally understand, and work with, the intricacies of the local ecology as they time the movement of their animals for the benefit of both animals and plants.

In fact, a "new" model of management has been developed to capitalize on such traditional systems in arid environments. Called "opportunistic management," it is often very similar to the herding practices of many indigenous pastoralists in arid regions. It is also strikingly similar to the traditional land use systems of the North Africans before French conquest in the nineteenth century. This predominantly pastoral population kept a wide variety of livestock, and much of it was managed by nomads and seminomads under conditions of high mobility appropriate to the ecology of the region. The more sedentary parts of the population were, by and large, agropastoral, and they raised crops in addition to livestock. Fire was often used in a sustainable manner to clear and prepare agricultural land and to create lush pastures.

As the following chapters demonstrate, however, the positive aspects of traditional land use in the Maghreb were not appreciated by the French in North Africa. Rather, the declensionist colonial environmental narrative facilitated the expulsion of the indigenous populations from their best lands to make way for colonial agriculture. The process not only systematically disadvantaged the North Africans but also led to profound changes in the landscape, some of which produced land degradation that continues to plague the Maghreb today.

Chapter 2

Nature, Empire, and Narrative Origins, 1830–48

NORTH AFRICA WAS WELL KNOWN to many in France, and in Europe more generally, long before the conquest of Algiers in July 1830. Travelers' accounts, missionary stories, consul reports, scholarly readings of the Greek and Roman classics, and commercial contacts had all provided various forms of information about the land and the people of the Maghreb for several hundred years.[1] The general consensus was one of admiration for the great fertility of the soil and the beauty of the landscape, accompanied by a rather ambivalent attitude toward the Algerian peoples. This was a very different view of the local people than the overwhelmingly negative one evident just a few years later in French writings on Algeria under occupation. By the 1840s, this new colonial narrative of environmental destruction by the "natives" was already influencing laws and policies in Algeria in significant ways.

Preconquest Views of the Maghreb

One of the oldest and most deeply ingrained images of North Africa held by Europeans was of its spectacular natural fertility. Descriptions of the fecundity of the soil, the abundance of agricultural harvests, and the thickness of the forests were first found in the writings of ancient Greeks such as the historian Herodotus (484–25 BCE), and ancient Romans, including the geographer Strabo (64 BCE–20 CE), Pliny the Elder (23–79 CE), who wrote the frequently cited *Natural History,* and the geographer Ptolemy (90–168 CE). Many of these authors recounted what earlier writers had said about places like North Africa without ever having traveled to such locations themselves. In the first chapter of the fifth book of *Natural History,* for example, Pliny cites the Greek historian Polybius as one of his sources confirming that near Mount Atlas "there are forests teeming with wild animals."[2] Pliny also relies on the Roman consul Suetonius Paulinus for his description of the regions around the Atlas as being "filled with dense and lofty forests of trees . . . with very tall trunks" and "neighboring forests [that] swarm with every kind of elephant."[3] These specific passages were often quoted by those French who wrote about North Africa. Another frequently cited ancient source was Strabo. Strabo also borrowed heavily from earlier authors in his monumental *Geography,* written for Roman imperial administrators. One of the most frequently cited passages from ancient writers was Strabo's declaration that "all of [Africa] situated between Carthage and the columns of Hercules (from Tunis to the ocean) is of an extreme fertility."[4] In the century leading up to the French conquest of Algeria, these ancient sources provided a significant proportion of not only citations but also basic information found in the writings of French, British, and other European authors on the Maghreb.[5]

One of the most influential of the eighteenth-century authors on the Maghreb, Thomas Shaw, the British chaplain in Algiers for twelve years, published his *Travels or Observations Relating to Several Parts of Barbary and the Levant* in 1738.[6] It was translated into French in 1743, and revised in a second English edition in 1757. His book is replete with references to authors of the ancient world, and he repeats many of their descriptions of fertility and agricultural abundance alongside his own observations.[7] This book, and another written by an American author, William Shaler, are two of the most frequently cited sources in French writing on the Maghreb in the period leading up to and immediately following the occupation of Algiers. Shaler, an American consul at Algiers, published his book in 1826; it was translated into French in 1830.[8] These two works, along with the two-volume study by the Abbé G. T. Raynal, constitute some of the most important sources consulted by the French in the early years of the Algerian

Figure 2.1. A map of Roman Algeria. This map was placed at the beginning of the section devoted to Algeria in Thomas Shaw's eighteenth-century book. He was one of the first authors to include maps of Roman North Africa in his writing on the contemporary Maghreb. Shaw included a similar map depicting Roman Tunisia at the beginning of his section on Tunisia. From Thomas Shaw, *Travels or Observations Relating to Several Parts of Barbary and the Levant* (Oxford: Printed at the Theatre, 1738). Reproduced with the kind permission of the Harry Ransom Humanities Research Center, University of Texas at Austin, Austin, Texas.

occupation and the years that preceded it.[9] Raynal's work was not based on personal experience traveling in North Africa but was derived instead from earlier sources, including many of the ancient classics as well as more contemporary works.

Two other works by Frenchmen must be added to this list. Although they were not published until 1838, and so were not widely read until then, they were circulated in manuscript form among many of those most concerned with the Maghreb well before the conquest of Algiers. After their publication they were cited quite frequently, if selectively. The first was written by Jean-André Peyssonnel, a physician and amateur natural historian who traveled in the coastal areas of North Africa in 1724 and 1725.[10] The second was by René Louiche Desfontaines, a botanist who undertook a scientific expedition to Tunisia and Algeria in 1785 and 1786. Although many of his writings were not published until the 1838 volume assembled by Dureau de la Malle (published together with Peyssonnel's book), Desfontaines did publish several articles during his lifetime.[11]

These and other books of the period provide numerous descriptions of the natural fertility of the soil in Algeria, references to the various agricultural, mineral, and other products to be found there, and historical overviews and lively descriptions of the local peoples. Although the descriptions of the agricultural fertility of North Africa found in these works are very similar, and in some cases identical, to those written after the conquest of Algiers, they differ dramatically in many other respects. The natural environment in these books is not portrayed as a degraded, deforested, and ruined environment, as it would be within just a few years of French administration. The local populations, and especially the nomads, are not described as destroyers of the environment, nor are they portrayed historically as the invading Arab bedouin hordes who ruined North Africa in the eleventh century. The attitude toward the local North Africans, though not always flattering, was much more nuanced and complex before colonization began.

The natural environment was most often described by these early authors in terms of the abundance of its agricultural production. Since most early accounts were limited to coastal areas of North Africa, the most common crops mentioned were grains, such as wheat and barley, some vegetables, as well as fruit and olive orchards in many areas. Accompanying many of these reports of agricultural abundance were descriptions of the plentiful water found in many areas. In 1738 Shaw observed that "the country round about Shershell [Cherchell] is of the utmost fertility, and exceedingly well watered by the Nassara, Billack, and Hasham" rivers.[12] Similarly, Shaler mentioned nearly a century later that "the soil in this part of Africa has lost none of its fecundity so famed in the past. . . . [It] is a country well watered, abundant in springs."[13] Shaler drew heavily from Shaw's work and cited him frequently throughout his book.

The French authors of this period were likewise eloquent in their descriptions of the fertility of the Maghreb. Abbé Raynal, for example, echoing Strabo, exclaimed that "all the Northern coast of this part of Africa . . . especially the plains which are between the sea and the Grand Atlas, are of a surprising fertility in wheat, in barley, in pastures, in vegetables, in fruits."[14] Raynal added that "the verdure and the vegetation there are maintained by abundant waters despite the heat and the drought of the summers."[15] Peyssonnel, too, took note of the fertility of the land and the good cereal harvests in several places.[16] Desfontaines bordered on the lyrical in his descriptions of "this superb country [Algeria] . . . [where] the plants renew themselves ceaselessly during all the seasons of the year, and [where] they reap the most beautiful harvests in all of Barbary."[17] He even claimed that "the country is so fertile that the Algerians do not want it to be visited by Christians in the fear that it will become an object of conquest."[18]

Forests and trees other than fruit trees receive little attention in most of these accounts. Importantly, though, there are no descriptions of deforestation or overgrazing by nomadic or other livestock. When forests are mentioned it is usually in connection with agriculture. Raynal briefly mentioned forests when describing the seasons. Shaw explained that the mountains in the region of Cherchell are cultivated up to their summits with fruit trees.[19] M. Renaudot, who had been an officer at the French consul in Algiers, described the mountains in the kingdom of Algiers as being "planted in vines and forests in nearly all their range."[20] An exception to this trend is the botanist Desfontaines, who did describe many forested areas of the Maghreb in detail.[21] He emphasized the favorable forest conditions and did not describe deforestation or reckless burning by the local populations. Generally, though, neither forests nor forest products seem to have interested these early authors. Their focus was on agricultural and other products like minerals that might profitably find markets. Forest products are not named in the lists of exports from the Maghreb countries during this period.

Many of these accounts of fertility were closely followed by explanations of the sad fact that the land was not producing up to its potential because it was not being husbanded correctly.[22] This was variously attributed to the terrible rule of the Turkish administration and/or to the laziness and ignorance of the agricultural populations. Shaler lamented that the despotism of the Turkish administration had limited the population growth of the kingdom of Algiers despite "a nice climate and a fertile soil."[23] Desfontaines remarked that "it seems that these countries were more fertile in the past than they are today, doubtless because they were cultivated with greater care."[24] Raynal explained that, despite the natural fertility of the soil and the abundance of the waters, "agriculture must be in the last stage of neglect."[25] Renaudot, who published the revised fourth edition of his book on the eve of the 1830 invasion of Algeria, wrote concisely that "the soil, although neglected, is of a surprising fertility."[26] He had earlier asserted that, under the crushing yoke of the Turks, "corroded by misery, burdened by harm, dying of hunger on the most fortunate and the most fecund soil, they [the Algerians] scorn cultivating the treasures that nature has brought them."[27] Importantly, though, none of these accounts report wanton deforestation, overgrazing, or environmentally destructive habits among the local populations.

Still, the attitudes toward the local populations evidenced in these early works were varied and complex in comparison to the simplistic accounts that were developed later under colonial administration.[28] Most often the population was classified into three or four groups. The Turks were always identified as a separate group, as was the Jewish population. The non-Turkish, non-Jewish population was frequently identified as Maures, or broken down further into separate groups for Maures, Arabs, and sometimes Berbers (Kabyles in Algeria). In this case, the Maures were defined as city dwellers descended from the Muslims who had conquered and then fled Spain; the Arabs were defined essentially as rural people, including nomads; the Kabyles were those living in the mountains.[29] Some authors such as Raynal also mentioned the Christians and the "blacks."[30] The Turks were universally described as tyrants, barbarians, and oppressors, and the Jews were most often portrayed as businesspeople, moneylenders, and sometimes as executioners. The urban Maures were also categorized largely as businesspeople but also as soldiers and farmers. The Berbers, always defined as mountain-dwelling, aroused hopeful curiosity or dislike, depending on the author. The large segment of the population that lived in the plains of the Maghreb was sometimes called the Maures of the countryside, or simply Arabs. This group formed the majority of the farmers and herders. The nomads were frequently referred to as the Arab bedouin.

Because the non-Berber rural population, and especially nomads, were largely vilified during colonial administration, it is important to understand that they were portrayed very differently before the conquest of Algeria. As explained above, the farmers were generally described as ignorant and lazy, but not as inherently bad or destructive people. Descriptions of their husbandry did not include wanton destruction by cutting down trees or malicious burning. Renaudot explained, for example, that at the beginning of the fall agricultural season, "the Maures, to fertilize everything, set fire to the weeds and thistles that cover the fields they want to plant."[31] This seemed to him a perfectly reasonable part of the agricultural cycle, the rest of which he also described in a positive tone. Raynal provided a similar description.[32]

The nomadic pastoral population was likewise largely portrayed as inefficient but not as inherently bad. Renaudot wrote that "the plants and lentisk that cover the immense plains serve to nourish the herds, which are not even as numerous as they could and should be."[33] He noted that some nomads planted crops and then moved on to pasture their flocks in the desert while the crops matured. Shaler added that "the Arabs live in the plains of this kingdom: they live under tents, and continually change the place of their residence, according to the seasons and the abundance of the pastures. They have the customs of shepherds, and, probably also, the vices and virtues of their primitive ancestors."[34] Raynal remarked that "these peoples have conserved with a singular consistency the primitive simplicity of their habits and their clothes. . . . The unique occupation of these Arabs is reduced to the care of their herds, of their horses, to hunting and to war."[35] Raynal's comments here and elsewhere echo certain European attitudes of the period, which tended to idealize the "noble savage"—in this case, the "noble bedouin."[36] This is found again in Desfontaine's comment on the "primitive purity among the bedouins."[37] Peyssonnel described the Arab nomads as poor and miserable but explained that they "are hospitable and charitable; they receive without distinction all who demand of them hospitality, Muslims or Christians."[38]

Raynal provided one of the clearest distinctions between the Arab bedouin, who tended their flocks and fought only when necessary, and a "few other small tribes who have neither tents nor stubble . . . [who] wander the desert looking to plunder; they are called with reason *Arab thieves.*"[39] Even this negative description, though, he tempered with the cautious comment that "truthfully they don't intend to kill, and they do not kill when they are not resisted, and when one does not refuse to pay the tribute which they demand."[40] Many of the other authors of the period acknowledged that some bands of Arabs, but not the majority, were forced out of necessity to become thieves. Renaudot and Shaw gave their own descriptions of the generous hospitality of many of the Arab nomads they encountered.[41] The overall view of the nomadic population, then, was mixed, but not universally bad. Importantly, their husbandry practices were not portrayed as harmful or environmentally destructive, only as primitive and unproductive.

A crucial part of the narrative developed during the colonial period was a history of North Africa that portrayed the spread of Islam across the Maghreb with a singular focus on the destructive eleventh-century "Arab invasion" from the East. By this account, the primary agents in the ruin of civilization *and* the environment were the hordes of Arab nomads who had sacked and burned cities, cut down all the trees, and, with their hungry herds, overgrazed the land. In contrast, few of the preconquest sources even mention the eleventh-century migration of Arabs across North Africa. Instead, the singularly disastrous event according to the majority of the sources available before 1830 was the fifth-century Vandal invasion. Peyssonnel, for example, considered the invasion of the Vandals to be one of the most noteworthy events since the Roman era, because it was "under them, principally, that the cruel persecutions of the Christians arrived."[42] A century later, Raynal agreed, adding that "no part of this vast and fertile province [North Africa] escaped their destructive rage. They uprooted the vines, cut down the trees, destroyed the harvests."[43] In these early texts, discussion of the spread of Islam focused on the seventh-century arrival of the Arabs from Arabia. In some cases the authors deplore the subsequent attack on North African Christendom by Islam, which is a long-established theme in France by this time.[44] The most common descriptions, however, were of the quick passage of the Arabs across the Maghreb and of their strong governments both in North Africa and in Spain. Peyssonnel did note that the Arabs "destroyed and razed all the cities [and] burned all the buildings."[45] Desfontaines also remarked that the seventh-century Arabs "ravaged the majority of the cities," but added that in some cases cities were "destroyed and rebuilt by the Arabs."[46] Even the accounts of destruction, which were not universal, were limited to urban areas, and there were no descriptions of environmental destruction by invading Arabs in the early texts examined here.

The big picture that emerges from the preconquest European texts on the Maghreb, then, is one of a fertile region, not living up to its potential though full of promise, which was in need of liberation from the oppressing Turks. It was primarily this oppression, mixed with ignorance and apathy, that was said to prevent the proper husbanding of land that could produce more and better products. The environment was described primarily in terms of its abundant agricultural and mineral products; forests, when mentioned, were described as thick and generally to be found in or near mountains with plentiful streams. The Arabs, especially the nomads with their large herds, were portrayed as mostly hospitable but indolent and lazy. The attitudes such descriptions helped generate in France are nicely summed up in the words of Renaudot, written on the eve of the 1830 expedition to conquer Algiers: "[B]ehold a superb country and great riches to conquer, a people to deliver from servitude, and Europe [to deliver] from piracy."[47]

Trade, Politics, and History

In the years leading up to the 1830 conquest of Algeria, piracy in the Mediterranean Sea received greater popular attention than the small bands of "wild Arab" thieves in the interior of the Maghreb. The need to liberate Europe from these terrible pirates formed one of the many justifications for attacking Algiers. "Take up arms against these tyrants of the sea," Renaudot exhorted his countrymen, "destroy their government, burn and annihilate those who contributed to maintaining their savage despotism."[48] Pirates could and did affect the trade carried out in the Mediterranean, although their influence had declined considerably during the eighteenth century. France, in particular, had well-developed trade relations with various regencies in North Africa, including on-site factories, and commercial relations can be traced back to the thirteenth century.[49] Commercial relations between France and the Maghreb had become quite extensive and profitable by the closing years of the eighteenth century despite continued claims of problems with piracy.

In Paris a more immediate concern had the attention of the French government in the late 1820s. The popularity of King Charles X, the last of the Bourbons, was waning precipitously.[50] The Bourbons had been restored to power following the fall of Napoléon I. The Revolution had not only ushered in more representative government and a constitutional monarchy but also paved the way for capitalist and industrial development during the nineteenth century. Although the kings of the Bourbon Restoration initially honored many of the reforms brought about by the French Revolution, their rule became increasingly repressive and unpopular. By 1830, the government was searching for ways to increase the king's popularity and to prevent yet another revolution. Based on an exaggerated claim of the need to end piracy in the Mediterranean, as well as exaggerated offense taken over an "insult to French honor" by the dey of Algiers, the French government announced in the spring of 1830 that it would attack Algiers. In late May, 635 ships carrying nearly forty thousand men attacked at Sidi Ferruch just outside Algiers. On 5 July, the French claimed victory over the dey and took possession of Algiers.[51] Although largely approved of by the French public, the conquest of Algiers did not restore the popularity of the king. A three-day revolution ensued in July after elections unfavorable to the king had been declared invalid. The Duke of Orléans, King Louis-Philippe, was placed on the throne on

30 July as the country's new "citizen king" of what became known as the July Monarchy.

Many scholars have argued, however, that the reason for the invasion of Algiers was at root economic. A large debt had been incurred by the French government at the turn of the nineteenth century by the purchase of Algerian grain for Napoléon's campaigns in Egypt and elsewhere. Estimates of the debt to Algerian merchants range from seven to eight million francs. Napoléon I never repaid this debt during his rule from 1799 to 1814.[52] The Restoration government had little interest in paying the debts of a previous government to a foreign power. In 1827 the French consul at Algiers was pressed by the dey to repay the nearly thirty-year-old loan. Reportedly, the consul refused to discuss the matter and the dey hit him with a fly whisk and dismissed him.[53] This was the insult to French honor mentioned above; France broke off diplomatic relations and within a few weeks had declared a blockade of the Algerian coast. The dey retaliated by destroying some French trading posts in his territory. The situation continued to deteriorate, ultimately leading to the French attack on Algiers in 1830.

Other factors, however, were also at play in this complicated political and economic ballet. As a result of the Napoleonic Wars, by 1815 France had been deprived of its overseas territories and was left with no colonial possessions.[54] Algeria would be the first territory newly conquered and occupied by France in nearly thirty years, since France vanquished and quickly lost Egypt at the turn of the century. In addition to sources of raw materials and agricultural products such as cotton, coffee, and indigo, the French government was also, in the wake of a series of revolutions, looking for a place to relocate segments of the population deemed undesirable or dangerous. It was taken to task for this latter goal by some, including a former member of the scientific commission of Algeria, Prosper Enfantin, who warned in 1843 that it would be monstrous to do as some had suggested and "make Algeria the Botany Bay of France . . . discharging in Africa, as in a sewer, the defective and the destitute of the population."[55] He added that "this would be a blundering imitation of England, because, thank God, the indigenous population of Algeria merit more regard and respect than some cretinous savages of New Holland [Australia]."[56] Indeed, the French government had been pondering actions such as these quite early in the occupation, as revealed in the 1833 report of the African Commission (Commission d'Afrique). This commission, sent by the government to assess the Algerian situation after the conquest, concluded that "economic calculations had belittled the value of colonies. The old nations must have outlets in order to alleviate the demographic pressures exerted on big cities and the use of capital that has been concentrated there."[57] The commission recommended that "to open new sources of production is, in effect, the surest means of neutralizing this concentration without upsetting the social order. . . . It is the surest way of preventing the seeds of hostility that are being sown among working classes, not only against the government but also against society and against property."[58] Several years later, the government would also be taking into account the value of new export markets in the colonies for the growing number of products being manufactured in France.

A further impact of the Napoleonic Wars was the exacerbation of an economic and political decline in Ottoman Algeria during the late eighteenth and early nineteenth centuries. For decades, Algeria's primary trade relations had been with France. Napoléon's invasion of Egypt in 1798 disrupted diplomatic and commercial relations between France and Ottoman Algeria.[59] A temporary blockade caused Algeria to lose many important markets for its products. European grain purchasers, for example, switched to suppliers in Russia, and Algeria regained only a small portion of its previous markets after the blockade was lifted. These commercial disruptions had significant impacts in Algeria. In addition, beginning in about 1805, the government in Algeria experienced revolts by the populations in the hinterlands that taxed its resources. The Ottoman government of Algeria thus was weakened both politically and economically when the French attacked.

Economic decline also contributed to urban depopulation. It is estimated that the population of Algiers was reduced from a hundred thousand in the eighteenth century to only about thirty thousand at the time of French occupation.[60] Another important consequence of this commercial rupture and economic decline was the reduction of cereal cultivation due to depressed prices and limited markets. Land increasingly came to be used instead for grazing in many areas, for example around Mitidja, Bône, and Arzew.[61] The society the French found in Algeria in 1830 was therefore less urban, less agricultural, and more pastoral than it had been in many years. This likely heightened the common perception among the French that this was an empty, unused, wasted land waiting to be brought under enlightened production.

Tending the Land with Fire:
Nature and Culture in the Precolonial Maghreb

More than anything else, it was the traditional use of fire that convinced the French that the Algerians were ignorant, destructive, and rebellious. Where they weren't burning the forests, the "anarchic overgrazing" by their livestock was said to be finishing off the forest and other desirable vegetation. Most of the French simply did not conceive of the forest as a place to live and use. Certainly most French administrators, settlers, foresters, and capitalists did not see forests as anything other than sources of timber, cork, and other forest products. Somewhere between a quarter and a half of the rural Algerian population, however, depending on the region, was dependent on forest products and/or living in forested areas in the early years of French occupation.[62] The other half to three-quarters of the population was composed of nomads or seminomads who were mobile and raised large herds of sheep, goats, and camels. This way of life did not appear any more "civilized" or rational to the French than did living in the forests. A fundamental clash of perceptions and opinions regarding both forests and pastures, then, gave rise to two of the most important and long-lasting points of contention, and areas of repression, found in the entire colonial period.

The most widely accepted estimate of the Algerian population at the time of occupation is approximately three million people, only 5 percent of whom lived in urban areas.[63] The remaining 95 percent lived in the mountains, plains, and deserts, where they survived by raising livestock and farming. More than half of Algerians, in fact, were nomads or seminomads. Estimates of the nomadic population range from 60 to 65 percent of the total population at the time of occupation.[64]

Algeria was, before French occupation, fully self-sufficient in agricultural products and even exported surpluses, especially of grains. In the mountains, fruit trees provided many of the primary crops, and some grains were grown. The plains were largely given over to cereal cultivation, mainly wheat and barley; vegetables were also grown, in many areas with irrigation. In the deserts, livestock husbandry predominated, except in the oases, where dates and vegetables provided the majority of agricultural products. In nearly all the rural areas livestock were raised in numbers and combinations that varied with the region, the people, and the environment.

One of the most frequently used traditional land management techniques was semicontrolled burning, used to improve production for both traditional pillars of the agricultural economy: grains and livestock. In unforested agricultural areas, existing vegetation was burned prior to planting cereals.[65] The ash of these stubble or fallow plants provided a rich fertilizer that often precluded the need to add manure, and burning also reduced weeds and pests. Burning usually occurred toward the end of the summer, and planting followed in late summer or early fall. This agricultural cycle was adapted to local precipitation patterns, which include dry summers and wet winters. Wheat and barley were harvested in the spring. Land in the forested areas was also tended by burning. Fires were set to clear land for planting and to stimulate the growth of pasture plants for livestock. Called *ksir* in Arabic, this technique generally followed a four-year cycle in which burning, planting, and harvesting occurred during the first year and the land lay fallow for the following three years. It was used to prepare agricultural land in forested areas not only to provide fertilizing ash but also to clear shading vegetation and thus increase the sunlight available for crop growth. The ash and sunlight gained produced good pasture in many forest areas where livestock were grazed. The fires also tended to drive away predators, such as panthers, from the parts of the forests that people were using, thereby keeping livestock and people safer.[66] Other quite sophisticated land management techniques were also used in Algeria prior to French rule, including legume rotation, irrigation, and soil conservation methods, although some of these were lost during the colonial period.[67]

Lands not normally tended with fire included the extensive pasture lands south of the primarily agricultural Tell, those of the High Plateaus, and the desert south (see appendix, page 179). Nor was burning a common land management technique in desert oases, where careful management of irrigation made bountiful harvests of dates and vegetables possible. The extensive pasture lands were the domain of the seminomads and the grand nomads, and they required little tending.[68] These extensive pasture lands constituted the largest part of Algerian territory, and half or more of the population lived in these arid areas with their livestock. Conservative estimates calculate the number of sheep in Algeria at the time of occupation at approximately eight million head, nearly three times the human population of the entire country.[69] Goats were likely more numerous than sheep, given their hardiness, and the number of camels was certainly significant, although estimates for either are not available. In these pasture lands, where rainfall was scarce and erratic, the simple movement of livestock (and people) formed the primary livelihood tool. Living mainly in tents constructed of goat hair, these nomads migrated largely in response to precipitation patterns. In the dry summers, the move was made to northerly pastures or to areas of higher elevation in the High Plateaus. During the cooler, more humid winters, there was a reverse migration toward the warmer desert or lowland areas.[70] Both the seminomads and grand nomads usually also grew grains in varying quantities during the winter months in their traditional territories. In some cases these cereal fields were tended until harvest and in others left to mature on their own, to be harvested when the group returned from grazing elsewhere. Not infrequently, the northerly summer migrations were made in connection with a "pasture contract" (*achaba*), under which the livestock were allowed to graze on the agricultural stubble or fallow land of large proprietors in exchange for the nomads' labor.[71]

In fact, contrary to what the French believed when they first occupied Algeria, many different forms of property and land tenure existed, including private property. Much, though not all, agricultural land was held as private property (*melk*), whereas nearly all pasture land and some cereal fields were considered collective property (*arsh*), which was "owned" by a lineage group with use rights determined and regulated by that group. Some land (*habous*), especially in urban areas, was held by religious institutions and could not be sold. The Ottoman government controlled large amounts of land (*beylick* and *mokhzen*), much of which was used by government officials as payment for government service or loyalty. Some of the government land was also let for sharecropping by the poor and landless, who were allowed to keep one-fifth of the harvest. A final category, dead land (*muwat*), was recognized in some areas; this was unproductive land that could be claimed by whoever cleared and cultivated it.

In the Tell region, most land was owned either by the government or by large private landholders, both of which employed large numbers of poor and landless Algerians as sharecroppers. In the mountainous areas most of the agricultural land was held as private property, except pastures, which were more often collective land. Forested areas were nominally the property of the government in many areas, but traditional usage rights of the local Algerians were nearly universally uncontested. In the oases, land was sometimes held privately. More commonly, though, those living in oases were under the protection of strong nomadic groups that exacted large shares of their produce in exchange for keeping other nomads at bay. Much of the rest of Algeria, primarily the arid and semiarid pasture lands, were considered collective property; the regulation and inheritance of these lands was under the jurisdiction of the nomads.[72]

The first impression of the French upon occupying Algeria, though, was that nearly all land was owned by the state and that private land was practically nonexistent. Collective lands were not well understood and not recognized during these early years. Until the 1840s, fallow fields and pasture lands were mostly thought to be unused or ruined—and, by further logic, thus unowned and available for taking.[73] Few of the French understood the complex ecology of arid lands and the sophisticated use of these expansive pastures by the indigenous pastoralists.

Extensive livestock production under arid and irregular climatic conditions requires large amounts of land and a high degree of herd mobility, since only small areas of land produce browse at any one time.[74] Plant growth under such conditions is largely dependent on rainfall. During dry periods vegetation can be very scarce, especially the annual grasses and other annual plants that constitute approximately 50 percent of the species present. These plants reproduce from seeds in a yearly cycle, rather than by vegetative growth, and thus exist during the dry season as belowground biomass (seeds) invisible in the landscape. In such annual grasslands, since little vegetation is present during the dry season, aboveground biomass is an inadequate indicator of ecological health. In this environment, thousands of seeds are often found per square meter of soil, forming rich seed banks that will germinate with the next rain. The richness and variety of annual plants are noticeable only if ecological assessments are conducted after significant rainfall, and this was rarely done during the colonial period. Much of the regional ecology, then, with plants having evolved under such conditions over thousands of years, is well adapted to drought and to grazing, especially under the extensively mobile management practices of the indigenous pastoralists.[75] The majority of the French, however, saw these arid pastures as overgrazed, ruined, and abandoned.

Forests, on the other hand, were more of a concern to the new French administration during its first two decades in Algeria than were pasture lands and the southern desert regions. Occupation proceeded slowly, and areas beyond limited coastal enclaves were not conquered and controlled until the mid-1840s. French forces did not reach far beyond the Tell until the 1850s (see map 2). Thus, the French came into greater contact with people living in forests and agricultural regions than they did with extensive pastoralists. Unfortunately, most of the French did not understand forest ecology in arid and semiarid lands any better than they understood arid-lands range ecology. Nor did most appreciate the sophisticated and sustainable traditional uses of the forest by local peoples. They interpreted the local practice of ksir as inherently destructive, believing it produced a ruined, deforested landscape. However, just as much of the vegetation in North Africa is well adapted to drought and grazing, a large proportion of the vegetation, including many trees, is well adapted to fire. A majority of Mediterranean vegetation has coevolved with fire and regenerates quickly after burning.[76] In fact, many experts believe, in the words of one, that "fire is essential to most mediterraneoid [*sic*] ecosystems."[77] The numerous adaptations include the thick bark of the cork oak (*Quercus ilex*) with its fire-insulating properties, a trait that was appreciated by some late in the colonial period.[78] In any case, the relatively small, semicontrolled fires used by Algerians to prepare fields and enrich pastures in forests did not normally ignite large conflagrations. Much more destruction occurred, for example, in oak forests where the cork had been harvested, leaving the trees defenseless against fire. Traditional burning strategies often benefited the environment, contrary to colonial rhetoric vilifying them.

The occupying French military needed wood, and this focused the new administration's attention on Algeria's forests early in the colonial

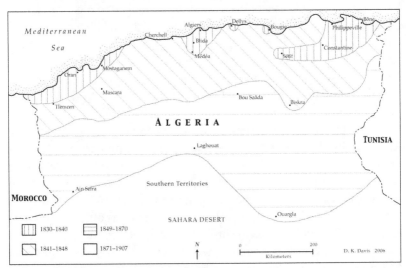

Map 2. Stages of occupation of Algeria. Modified after John Ruedy, *Modern Algeria: The Origins and Development of a Nation* (Bloomington: University of Indiana Press, 1992), 56.

period. The military forces were heavily dependent on forest products for their survival and functioning. Forests provided building materials, fuel, and some game where needed for the troops. The French administration was also very concerned with the acquisition of land, especially agricultural land, for the nascent but growing colonization efforts in the first territories conquered along the coast. It is in these sectors that some of the earliest transformations of the environmental narrative become evident.

Conquest, Occupation, and the Colonial Environmental Narrative

After the capture of Algiers in 1830, the conquest and occupation of the rest of the country proceeded haltingly. Despite an early victory at Algiers, the French government did not fully commit to colonial occupation until 1834.[79] Until this date, and indeed after, a fierce debate raged in France as to whether North Africa should be colonized or not, and no coherent policy of colonization was developed.[80] The colonialist lobby eventually won, and by 1838, the territories known as "the French possessions of North Africa" were renamed Algeria (Algérie). Until 1840, though, the official policy was one of restricted occupation, limited to the areas immediately surrounding major urban centers such as Algiers, Oran, and Bône. A large percentage of the country was not even nominally under French control. This area of occupation was enlarged to most rural areas of the Tell during the 1840s, primarily under the brutal direction of General Thomas Robert Bugeaud, governor-general of Algeria from 1840 to 1847.[81] Thus the coastal areas and the Tell, the areas with the best rainfall and the most vegetation, were not controlled and occupied for nearly twenty years. Much of the rest of the country, the more arid region, was not explored to any significant degree until well after this date.

Once the French had committed to the conquest and occupation of Algeria, they also committed to colonization. The military and the new French administration in Algeria needed to provide land, especially agricultural land, free or at low cost to encourage immigration. Much of this was accomplished early in the occupation by simple military force that drove many Algerians from their lands. In other areas other means were employed. One of these was the utilization of an environmental narrative fundamentally different from the preconquest narrative. As early as 1834, and possibly earlier, the Algerians began to be blamed for ruining the land, and this accusation provided some in the military and the administration with a justification to expropriate it.

One of the earliest examples of this transformation of the narrative is found during the pacification of Constantine Province. In 1834, the leader of a cavalry troop in Bône explained that, to obtain land from the Algerians, all that needed to be said to them was, "during all the time that you have occupied the soil, it [the soil] has been *destroyed because you are nomads*. If we want to we can take your lands."[82] This cavalry officer, M. Poinçot, concluded in his report to the council of ministers that "with similar arguments, it is nearly certain that we will obtain from the nomads the surrender of all the land we need at very low cost."[83] Seven years later, twenty thousand hectares of land were in European hands and accumulation was proceeding quickly.[84] By 1845, a model village at Ténès had been constructed by the local Arab Bureau, the primary purpose of which was "the progressive fixation of the Arab people [nomads] to the soil" in order to control them and their destructive way of life.[85]

Four years after Poinçot's report, two new government ordinances began to codify some of this revised narrative regarding indigenous land use. The first, passed 1 July 1838, forbade "the burning, for any cause, of woods, copse, brush, hedges . . . [and] standing grasses and plants" under pain of severe penalty.[86] The second, passed 18 July 1838, specified a large fine of fifty to one hundred francs for any incident of burning brush for the purposes of fertilizing the land.[87] By 1846, the penalty for burning in order to fertilize the land was raised to include from "six days to six months or more in jail."[88] These two ordinances were justified with claims of the inherent destructiveness of indigenous burning and of the need to protect the forests. They also effectively restricted one of the primary tools for working the land used by Algerians, thus making cultivation and grazing more difficult and time-consuming for most of the rural population.

One of the immediate motivations for protecting and preserving forests, in particular from the Algerians, was the military's need for wood during the first decades of occupation. Indeed this was one of the primary purposes of the new Forest Service (Service Forestier d'Algérie), which was established in 1838 under the direction of Victor Renou.[89] The Forest Service was charged with assuring the provision of wood needed by the seventy thousand French troops.[90] Civilian needs for wood were added to this, bringing the total necessary to approximately eighty to one hundred thousand hectares of forest per year.[91] The extent of Algerian forests was not known at the time due to the limited penetration of the country, and the director of the Forest Service had explored only eighty thousand hectares of forests by 1842.[92] During the period of unorganized and unrestricted colonization of the 1830s, Algerian forests suffered significant losses at the hands of the military and the French and European immigrants. Charles-Robert Ageron has graphically described this assault on Algerian resources as "a flight of human vultures . . . grabbing hold of land and cutting down the woods."[93]

The military was responsible for a great deal of damage during this early period of occupation. Beginning with the conquest of Algiers, the army felled trees regularly, including young trees and even cultivated olives when that was all they could find.[94] The battles of pacification also claimed many trees at the same time that they destroyed infrastructure and agriculture, and killed countless Algerians (see figure 2.2). One French military commander described the destruction of 1842 and 1843 in his letters: "[W]e ravage, burn, plunder, we destroy the houses and the trees. . . . What destruction! I saw millions of fig trees razed. . . . They [the Algerians] were occupied tending their wounded while I continued to burn their huts and cut down their fruit trees. . . . I killed the enemy, more than two hundred men, burned their huts, cut down their trees and their barley."[95] The need for wood was so great—and consumption in the early years proceeded so quickly—that the Forest Service began work on plantations and tree nurseries in the 1830s. By 1852, General Randon had even organized military tree planting companies (*planteurs militaires*) to plant areas deemed in need of reforestation.[96]

Commercial exploitation of the forests by Europeans did not commence, however, until about 1849. Since the administration did not understand or recognize traditional land tenure in Algeria during the first two or three decades of occupation, it claimed all the forested land found in the colony as the property of the state. As such, the French Forest Code of 1827 was nominally applied to forested areas from early in the occupation, although it was not recognized legally in Algeria until 1883.[97] Forests were formally declared the property of the state, with small areas held for indigenous use, by the law of 16 June 1851.[98] Large numbers of Algerians were thus expropriated of their forests. The regime of forest concessions (long-term leases for harvesting) formally began in 1846, and by 1861 some 102,000 hectares had been given to businesses in concessions and a further 145,000 hectares had been requested.[99] The state took from 10 to 30 percent of the profits of the forest products sold, primarily cork.

Perhaps because commercial exploitation of the forests did not begin until the late 1840s, and because the exploration of forested areas was cir-

Figure 2.2. *The Grottoes of Dahara,* 1845, T. Johannot. This image shows the burning to death of an estimated eight hundred to one thousand Algerians in a large cave where they were seeking refuge from the attacking French army in 1845. Such brutality was not uncommon in the early years of the French occupation of Algeria. From P. Christian, *L'Afrique française, l'empire de Maroc et les déserts de Sahara* (Paris: A. Barbier, 1846), facing 440. Reproduced with the kind permission of the Bibliothèque Nationale de France.

cumscribed until the 1850s, the negative environmental narrative obvious in Poinçot's report blaming the Algerians was not very evident in the writings of foresters during the first two decades of occupation. In fact, the director of the Algerian Forest Service, Victor Renou, held views of the forests more compatible with the preconquest environmental narrative.[100] In his best-known article, published in the *Annales Forestières* the year of his accidental death in 1844, he provided an interesting description of the cedar forests.[101] He extolled the forests' thickness and extent in mountainous regions and marveled at the enormous size of individual trees. He pondered their commercial value and considered the various ways to transport harvested trees. He evaluated the forest clearings and did not blame the local inhabitants, the Kabyles, for chopping or burning to create them. Instead he concluded that natural factors such as high elevation and steep slopes probably impeded tree growth. He included no discussion of any deforestation or environmental damage perpetrated by Algerians in these forests.

Early botanists in Algeria also maintained the relatively benign environmental narrative found in the preconquest period for most of the first two decades of occupation. Some of the earliest (preconquest) botanists and amateur natural historians, such as Desfontaines, Peyssonnel, and Shaw, as detailed above, had portrayed the flora and physical environment of Algeria in a positive way, stressing the natural fertility of the soil and the wide variety of rich plant life. They described the local peoples as ignorant of the best agricultural techniques without characterizing them as environmentally destructive. Until the publication of the botanical sections of the *Exploration scientifique de l'Algérie* in the 1850s and 1860s, however, new botanical research and writing was quite limited. One of the few publications that appeared was that of an Englishman who settled in Algiers in 1839, G. Munby. In the preface to his *Flore de l'Algérie,* he portrayed the vegetation and the natural environment in terms strikingly similar to the positive descriptions of the botanists who preceded him.[102] There is no mention of deforestation or environmental destruction in his work.[103] In the first two decades of occupation, then, the narrative did not change substantially among botanists and amateur natural historians.

Others who were neither foresters nor botanists also wrote about Algeria's plants and forests and thus contributed to the evolving colonial environmental narrative. Some of the most important of these early authors were members of the Commission Scientifique de l'Algérie. This group of twenty-one scholars, composed of eleven civilians and ten army officers, arrived in Algiers in December 1839 to carry out the scientific study of Algeria following the 1837 orders of the French government.[104] The resulting thirty-nine-volume *Exploration scientifique de l'Algérie* appeared over a period of twenty-three years, from 1844 to 1867, although most of the field work was carried out over the short period from 1840 to 1842.[105] Several of these volumes contain revealing descriptions of, and attitudes toward, the Algerian environment and local people, and thus provide early examples of how the environmental narrative was beginning to change under colonial administration. All of the members of the commission supported the occupation and colonization of Algeria, and their work, in the words of Michael Heffernan, "constructed an elaborate and intensely poetic image of North Africa and of France's future colonial rule in the region which was to exert a powerful long-term impact on the French imperial *mentalité.*"[106]

One of the most prolific of the commission members was army captain and geographer Ernest Carette.[107] The earliest of his volumes, appearing in 1848, treated the Kabylie, a mountainous area to the east of Algiers. In his description of the region, the people, and the economy, Carette marveled over the extent and thickness of the forests and the general fertility of the soil. Throughout the volume he described forests in great detail and included no descriptions of deforestation or environmentally destructive behavior by the Algerians. He concluded that the forests tended to grow according to soil patterns and on mountain summits. He was so impressed by them that he exclaimed in the forestry section that the Kabylie contained "an almost inexhaustible mine of wood," and left, therefore, "a large place for other consumers."[108] Carette seemed to share the preconquest view of the Algerian environment, at least in regard to this region, and indicated that there was potential for French exploitation of the Kabylie forests.

By contrast, other members of the commission highlighted the supposed loss of the ancient fertility and productiveness of the Algerian environment, echoing another aspect of the preconquest narrative. Army officer and renowned historian of North Africa H. F. J. E. Pellissier de Reynaud, in the first volume of the *Exploration scientifique* to be published, in 1844, described the numerous ruins as evidence of a dense population during the Roman period and thus of the high former fertility of the soil.[109] Like many at the time, he proclaimed that North Africa had been the granary of Rome. This inaccurate conception was strengthened by the prominent and widespread ruins of Roman towns, aqueducts, and cisterns, which seemed to indicate high populations and a plentiful water supply.[110] Pellissier de Reynaud also concluded that North Africa was now relatively devoid of forests due to deforestation that had occurred since the Roman period.[111] Although he stopped short of assigning blame for this deforestation, his description pointed to a change in the environmental narrative from one of an innate fertility not being properly husbanded to one of an ancient fertility that had declined or been destroyed.

This change in the narrative was further developed by another member of the commission, J. A. N. Périer, a civilian colonist and medical doctor. Périer wrote the two-volume section on the medical sciences (1847) for the *Exploration scientifique*.[112] In the first chapter, Périer surveyed the physical environment, climate, and vegetation as a prelude to discussing the possibilities of acclimatization of Europeans to Algeria. Périer repeated some of the earlier environmental narrative regarding the ancient fertility of Algeria, but he emphasized the decline of the environment since the Roman period. "This land, once the object of intensive cultivation, was neither deforested nor depopulated as today," he lamented; "it was the abundant granary of Rome."[113] He claimed that forests covered the country in the past and blamed Arab pastoralists and their "worship of [livestock] herds" for the subsequent deforestation. The proof, he claimed, lay in the fact that "the woods have been conserved everywhere where the Arab has not, for a long time, established his residence."[114] For Périer, woods, and the rainfall he thought they produced, were crucial to the overall healthfulness of a country. Périer provided one of the earliest, nearly complete iterations of what would become a ubiquitous story throughout the colonial period, and of the clear duty that was therefore held to fall to the French as the country's new stewards. As he put it, "it is our responsibility to raise Algeria from her fallen state, and to return her to her past [Roman] glory; for this privileged soil possesses all the elements of a surprising fecundity, of a great prosperity."[115]

Périer's account also highlights the fact that by the mid-nineteenth century, in the words of Patricia Lorcin, "the concept of the Roman legacy as integral to the Western tradition that bound Algeria to France had taken shape."[116] This is not surprising given that most educated people in France had been broadly trained in the Greek and Roman classics during the eighteenth century, and that many monarchists and Napoleonic imperialists worked to legitimate their regimes with allusions to the classical past.[117] The geographer Michael Heffernan has argued that at the turn of the century a "modern" France believed it "had a right and a duty to re-establish the traditions and values of the ancients in their former heartlands."[118] In newly conquered Algeria, a glorious Roman past was thus frequently invoked as both a justification for the conquest and a model to follow for successful occupation and colonization. As early as 1833 the king of France, for example, exhorted the expeditionary troops in Algeria to "finish the conquest and return to civilization this shore of the Mediterranean surrendered, since the destruction of the Roman Empire, to anarchy and barbarism."[119] In 1833 the French minister of war asked the prestigious Academy of Inscriptions and Literature to establish a history of Roman colonization in North Africa that might serve as a model for the French occupation of Algeria.[120] Although this endeavor resulted in limited work on the subject, it set into motion ideas and goals that would eventually be realized with the *Exploration scientifique de l'Algérie*.[121]

Many of the authors of the *Exploration scientifique de l'Algérie* expanded the goals and duties of the French in Algeria, based on the Roman model, to include the restoration of the mythical fertile environment of classical antiquity. Some, like Périer, provided suggestions for how to begin to restore the allegedly ruined environment. He agreed with the laws already established banning the burning and cutting of the forests by Algerians and added that "the disorderly grazing of herds" must also be curtailed. If these destructive indigenous practices could be stopped, he claimed, the forests would regrow and Algeria would become a healthier, more productive place.[122] Périer thus echoed the sentiments of the officer Poinçot, who blamed Arab nomads for ruining the environment and thus justified the expropriation of their land.

This early sentiment against nomads was also evident in one of the first official treatises on Algerian agriculture. The author, Professor L. Moll, member of the Royal and Central Agricultural Society, was sent by the French minister of war in 1842 as the first expert to study Algerian agriculture and to suggest the best methods to ensure successful colonial agriculture. Moll was clear about what needed to be done with the Arab nomads. He stated that "the nomadic life of the Arabs is an obstacle to our domination as well as to the civilizing action we must exercise. It would be good, I believe, to encourage as much as possible the establishment of fixed abodes."[123] He blamed the Arabs for deforestation, which he claimed had been caused by their livestock, their agricultural practices, and the fires they set. In a particularly vitriolic passage he exclaimed that in the "twelve hundred years [since] the Arabs invaded Africa . . . their rage of destruction has succeeded in transforming the country. . . and today we must initiate agriculture to rediscover, under the veil of desolation which covers [Algeria], the immense resources which this land of Africa still harbors."[124] Moll went further than most others, however, suggesting that the Algerians were so backwards and destructive that they should become extinct. In his opinion, "the entire earth rightfully belongs to civilization, and each race which is not fit must necessarily disappear as have disappeared the antediluvian animals."[125] Sentiments similar to these would become increasingly common in Algeria as colonial administration progressed.

By the mid-to-late 1840s, then, a new environmental narrative had clearly begun to emerge within certain segments of the colonial apparatus concerned with Algeria. It appears to have been used mainly by those with an interest in appropriating resources (land, forests, pastures) or controlling populations deemed dangerous (nomads), that is, by those promoting colonization and the military and administration. The two ordinances of 1838 banning the burning of woods and vegetation provide a good example of laws that were justified on the grounds that indigenous practices ruined the environment, especially the forest. The rationale underpinning these laws also reveals a profound misunderstanding of the local environment and indigenous land use by the French administration.

The years from 1830 to 1848 constituted a period of nearly completely unregulated and unrestricted colonization in the areas of Algeria under French control. During these first two decades of occupation and colonization, the administration placed a severe economic squeeze on the local population, which disempowered and disenfranchised them. In addition to the appropriation of many of their lands and traditional resources such as forest products, the Algerians were forced, sometimes brutally, to pay special taxes (*impôts Arabes*) and to submit to French judicial authority. Under the administration of Governor-General Bugeaud, two key laws passed in 1845 further impoverished the indigenous populations. The first was the law of finances of January 1845, which reformed much of the administration's financial system and, importantly, decreed that the Arab taxes must be collected in cash rather than in kind, as they had been up to that point.[126] This enmeshed most of the Algerians in the money economy, forcing some to sell land or other belongings and others into wage employment in capitalist agricultural ventures or in the cities. The second law was the ordinance of April 1845, which divided Algeria into three provinces (Oran, Alger, Constantine) and three different types of territories. The areas where all the population was composed of indigenous Algerians, called the "Arab territories," were subject to complete military administration. The areas where a considerable majority of the population was European, the "civil territories," were allowed a large degree of self-government and French common law. The "mixed territories" were governed by the military, with limited self-government for European residents.[127]

These first two decades of occupation witnessed rapid immigration and colonization by Europeans and the dispossession of Algerians of their lands and livelihoods in multiple ways. By 1851, for example, approximately 151,000 immigrants had settled in 176 centers of colonization, with 33,000 in rural areas, and more than 400,000 hectares of land had been allocated to settlers by various means.[128] Some land had been abandoned as Algerians were killed or fled before the occupying military forces. Some land was confiscated as punishment from Algerians who rebelled against the French, or was simply appropriated by the administration in the name of the public good. Some land was purchased at absurdly low prices from Algerians whose livelihoods had been devastated by the occupation. The lands of the Ottoman Turkish administration were declared property of the state.

The land thus obtained was not enough, however, to keep pace with the immigration promoted by the administration, especially under Governor-General Bugeaud.

In April 1841 Bugeaud passed one of the first colonization decrees, giving land grants of between four and twelve hectares to each immigrant having a minimum amount of capital with which to make improvements.[129] This was followed in 1844 by an ordinance governing *habous* (religious) and "vacant" land, and in 1846 by another governing the verification of land titles.[130] Together, these two laws resulted in the confiscation by the state of hundreds of thousands of hectares of land, especially land deemed vacant, such as grazing lands or fallow fields. Around the city of Algiers and in its hinterland, for example, 200,000 hectares were seized under these laws; 168,000 hectares went to European colonists and only 32,000 (16 percent) were left for Algerians.[131] Much of this seized land was communal pasture land, which, since it was not cultivated according to European methods, was defined as unused or vacant. Whether the rationale for these laws rested solely on a misunderstanding of indigenous land use or also, to whatever degree, on the newly evolving environmental narrative blaming Algerians, especially the Arab nomads, for destructive land use, is not clear. Given its growing visibility in publications such as the *Exploration scientifique de l'Algérie* and Moll's influential treatise on Algerian agriculture, however, it is likely that the new declensionist narrative played a role in these shifting political attitudes toward land ownership and use.

Yet this narrative was neither ubiquitous nor monolithic by 1848. In this early period, permutations of the precolonial and colonial environmental narratives were used by various actors for different purposes. Some in the military administration, like Poinçot, invoked the declensionist narrative to control nomads and to acquire land for colonization. Others in the military, however, did not. One officer, for example, wrote in 1830 that "this land, which has been presented to us as savage and uninhabited ... is covered in pretty houses ... [and] the vegetation is superb."[132] Nor, as late as 1847, did officers Ernest Carette and Auguste Warnier invoke the negative narrative in their official publication, even when describing the nomads, although Warnier later embraced the colonial narrative during his long and influential tenure in Algeria.[133]

Furthermore, though historians commonly argue that the colonialist lobby asserted the marvelous fertility of North Africa, while the anticolonialists rebutted that North Africa was an infertile desert waste, the contemporary arguments were far more complex. Some in the colonialist lobby, for example Adolphe Dureau de la Malle, highlighted the fertility and productive potential of the land, noting that much of it was vacant and available for colonization (see plate 2).[134] Others, especially in the 1840s, insisted that the innate fertility had been ruined by the Algerians and their destructive land use practices and that therefore their land should be taken and their way of life suppressed. This includes Professor L. Moll, whose rather extreme arguments are detailed above. Another example is found in the writings of a prominent and wealthy Algerian colonist, the comte de Raousset-Boulbon. This nobleman extolled the innate fertility of the soil but argued that the Arabs, "ignorant, vulgar, lazy, lying, bloodthirsty, cynical, degraded by all the vices" should be fixed to the soil.[135] He explained further that "the earth belongs to those who can, to those who develop it," and that "all of the soil of Algeria should . . . pass from the hand of the native to the hand of the European," but "only by expropriation."[136] In a collectively produced text, other prominent European colonists shared these sentiments and added that the Algerians "are workers of an inferior order, but . . . wholly acclimatized" to the hot climate, unlike imported European workers.[137]

Most in the anticolonialist lobby did not accept or use the precolonial environmental narrative of the Algerian soil's fertility and productivity, nor did they invoke the emerging colonial narrative of ruin and decline by the "natives." Rather, many dismissed North Africa in general as "a barren wasteland, filled with savage beasts and wild tribesmen."[138] Yet a closer reading of some of these authors reveals that, while they shared other anticolonialists' concerns over the cost of Algerian colonization to the French state, their understanding of the environment was fundamentally different. French provincial deputy Amédée Desjobert, for example, detailed the economic and military losses of the Algerian occupation and argued that the costs far outweighed the benefits, then added that "the idea of colonizing Algeria can only be born of a complete ignorance of the simple notions of political economy."[139] He dismissed the colonialist goal of emulating the granary of Rome, saying that "we have established that we cannot colonize as did the Greeks and Romans."[140] He pointed out that forage for livestock production, due to the nature of the climate and the soil, cost three times more to produce in Algeria than in France and that therefore meat produced by Europeans cost much more in Algeria than in France.[141] He made a similar argument regarding grain production. Rather than simplistically asserting that the Arabs had ruined the environment, though, Desjobert argued that "the ravenous droughts succeeded by the torrential rains" were the cause of "the nomadic state of the Arabs; they cannot sedentarize, for sedentary cultivation cannot nourish them."[142] He summed up the situation by saying that "the nomadic land use of the Arabs . . . is a necessity imposed by nature itself."[143]

An apparently lone figure within the colonial administration who did not believe that the Algerians had deforested Algeria was the agronomist Auguste Hardy. Hardy was the director of the government's nursery and later the director of the Experimental Garden (Jardin d'Essai) at Algiers. Hardy worked in these capacities for a quarter of a century, from 1841 to 1867, and published several works on Algerian agriculture.[144] Well placed in his position at Algiers, Hardy was familiar with the colonial environmental narrative and wrote that "it is often said, and repeated again every day, that if Algeria were forested, the rainfall would be more abundant and especially allotted more uniformly."[145] As one who conscientiously carried out numerous agricultural experiments and studied the physical environment carefully, however, Hardy thought differently. He explained instead that "the deforestation of Algeria is a natural consequence of its climate; it is caused more by the pernicious influence of the harmful winds and the poor seasonality of the rains than by the pasturing of animals and the fires set by herders, where the cause is so constantly sought."[146] Hardy did advocate, though, trying to tame the harsh Algerian environment, and he specifically championed the construction of thick, multilayered windbreaks by planting at least four successive rows of different kinds of trees in areas where the wind blew frequently and strongly, drying out soil and vegetation.[147] This idea of planting trees to try to stop scorching desert winds would come to be quite popular later in the colonial period.

While a few individuals, and several in the anticolonialist lobby, may have had a slightly better understanding of the Algerian environment and indigenous land use systems, most certainly did not respect or like the Algerians any more than did members of the colonialist lobby, who mostly reviled the Algerians. Many in the colonialist group openly discussed extermination of the local population as a way of facilitating the colonial agenda. As colonization proceeded, the anticolonialist voices were marginalized and the colonialist lobby came to dominate most discussion of Algeria. The colonialist lobby, though, was not homogenous; opinions differed, and the colonial environmental narrative acquired several different interpretations. A relatively small but influential group, the *"indigènophiles"* (native lovers), did admire much about Algerian life and champion certain indigenous rights. These men reached their height of power in the Arab Bureaus under the rule of Napoléon III in the 1850s and 1860s.

Although it would not be fully conquered and pacified for another decade, Algeria formally became part of France in 1848 as a result of the new French constitution written under the Second Republic. By this time, as this chapter has detailed, the declensionist colonial environmental narrative had been developed and had begun to influence law and policy formulation in Algeria. Over the course of Napoléon III's Second Empire, the narrative would undergo further permutations and would become even more widely entrenched.

Chapter 3

Idealism, Capitalism, and the Development of the Narrative, 1848–70

WHEN ALGERIA WAS OFFICIALLY MADE A PART of France in 1848, the French government definitively committed to colonization. The focus of debates surrounding Algeria changed from arguments for or against colonization to how best to administer and govern the territory. Over the next two decades, leading up to the Franco-Prussian War of 1870 and the Algerian Insurrection of 1871, a battle raged in Algeria and in France between those who championed European settlers' rights to land and resources and those who sought to protect the property and quality of life of the indigenous Algerians. The colonial environmental narrative, in a variety of permutations, was used by many in making their cases for these various causes. By the end of this period, despite some dissenting voices, the declensionist narrative became the dominant environmental narrative invoked. A particularly significant tactic was the selective translation and citation of medieval Arab authors on the Maghreb, such as Ibn Khaldoun, to detail and to validate the narrative.

Colonialist writers further refined the environmental narrative during this formative period and in doing so were strongly influenced by the theory and practice of French forestry. Key political events in France and Algeria at this time led to the transformation of agriculture and land use in Algeria. The idealism evident in the work of the Arab Bureaus (and that of other indigènophiles), supported for several years by the policies of Napoléon III's "Arab Kingdom," was crushed by 1870, as the settlers triumphed and capitalist production spread, changing the Algerian environment and the administration of the colony profoundly.

Revolution, Disenchantment, and the "Arab Kingdom"

The workers' uprising in Paris, the Revolution of 1848, was warmly welcomed by the majority of Europeans in Algeria, especially by the French settlers. These colonists saw the revolution as a chance to shake off military rule in Algeria and to implement their ideas of complete assimilation.[1] The new constitution of this short-lived Second Republic declared Algeria an integral part of France and its civil territories became three new French departments. For a few short years, the settlers in Algeria experimented with electing municipal councils and effecting broader self-government. The new French president, Louis-Napoléon Bonaparte, elected in 1848 to lead the Second Republic, was the nephew of Napoléon I. Louis-Napoléon, the "prince-president," staged a coup d'état in 1851 and was declared Emperor Napoléon III of the Second Empire in 1852. Napoléon III put an end to civilian rule in Algeria and reimposed military rule. After both the 1848 revolution and the 1851 coup, the government deported many French workers deemed troublesome to Algeria, where they provided cheap labor.

Napoléon III ruled France for two decades, until his defeat and capitulation in 1870 as a result of the disastrous Franco-Prussian War. During this time he promoted industrial capitalism and, to a certain degree, workers' rights. By encouraging what he saw as economic and social modernization, he sought to "close the era of revolution by satisfying the legitimate needs of the people."[2] There was relative economic prosperity in the early years of the empire, and the emperor developed a reputation for being a champion of the poor and enjoyed broader political support than had many of his predecessors. Although Napoléon III began with authoritarian rule, this was tempered as his reign progressed, and by the 1860s the use of force in his administration had declined. In addition, steps were taken to further political liberalization. Strikes were legalized, for example, and parliament was given greater power. The motivation for many of these changes was the desire of the emperor to stimulate economic development. In 1860, a commercial treaty was signed with Britain "designed to intensify competitive pressures and to force the pace of modernization."[3] Over the course of the two decades of Napoléon III's rule, living standards in France did improve, although perhaps not quickly enough. Worker agitation was renewed with the strikes of 1869 and 1870 and opposition to the empire grew in the cities.[4] As the emperor's power in urban areas was eroding, however, Prussia defeated Austria in 1866, and Napoléon decided, under pressure, to go to war. He was defeated by the Prussians and capitulated in September 1870, at which time the Third Republic was proclaimed and a provisional French government established.

Earlier, during the brief period of the Second Republic, Algeria suffered the "years of misery," which brought economic crisis, recurrent epidemics, locusts, lean harvests, and famine during the late 1840s. One result of these years of misery, and of the preceding two decades of conquest and occupation, was the reduction of the indigenous population from an estimated three million at the time of conquest to about 2.3 million in 1851.[5] As for the European settlers, the short period of hope and excitement arising from their increased autonomy under the Second Republic was extinguished when Louis-Napoléon reimposed military rule in 1851. The settlers would become more and more disenchanted with the Second Empire over the course of the next two decades, as Louis-Napoléon's ideas about Algeria developed and were translated into policy.

The "prince-president," later emperor, favored the liberal economic ideals prominent in much of mid-nineteenth-century Europe, and he encouraged private enterprise not only in France but also in Algeria. In 1851, the last year of the Second Republic, several important measures were passed regarding trade and commercial relations between France and Algeria. These included enactment of a new commercial law and the establishment of the Bank of Algeria.[6] The commercial law liberalized trade between France and Algeria and lifted all tariffs on Algerian agricultural products exported to France. Tariff relief raised Algerian exports to France and stimulated an increase in agricultural production, especially of grains.[7] When the Crimean War (1854–56) disrupted the flow of grain from Russia, one of France's chief suppliers at the time, prices soared. As a result, much of the land that had been used as pasture in northern Algeria was transformed, once again, into grainfields.[8] Cotton also enjoyed profitable markets during the 1850s and early 1860s, which led to expansion of its production. Rising profitability stimulated speculation in land as well as in grain and other agricultural products, including tobacco, and drew both major capital investment and corporations to Algeria. As a result, agricultural production began to be concentrated in large landholdings, both private and corporate, a trend that would intensify throughout the colonial period. The newly established Bank of Algeria provided the eagerly awaited and much needed credit necessary for many of these enterprises.

During the 1850s an additional 250,000 hectares of land came under the control of European settlers by various means. This took place primarily during the period of military rule, from 1852 to 1858, under the leadership of Governor-General Jacques Louis Randon. Under the land laws of 1844 and 1846 a policy known as cantonment (*cantonnement*), or delimitation, was used to further dispossess Algerians from their land. The argument ran that the Algerians, especially the nomadic pastoralists, could not "productively" use all the land they had; therefore, what was not being productively used was simply taken, leaving them with only the land deemed adequate for their needs by the colonial government.[9] This opened up new land for Europeans and drove many indigenous Algerians away from the centers of colonization, a process of *"refoulement"* welcomed by many settlers.

While this sort of specifically government-assisted colonization was spreading in Algeria, "corporate colonization," also called "private colonization," was likewise being encouraged, frequently organized and capitalized by large corporations such as the Compagnie Genevoise and the Société Générale Algérienne, among others. Large land holdings were given or sold very cheaply to corporations conducting a variety of enterprises, from settling immigrant families to producing grain or cotton, to harvesting cork and other forest products. Most of these projects took place on land expropriated from Algerians. Much of this activity occurred during the 1850s under the direction and through the facilitation of Governor-General Randon, who was determined to obtain "sufficient territorial resources for European colonization to progress rationally."[10] An ardent colonialist, Randon was a firm advocate of cantonment who believed that the Arabs had to accept "refoulement, which is in the nature of things."[11] The result was that a growing number of landless Algerians were left with little choice but to provide wage labor for European ventures in the cities and countryside.

Near the end of Randon's command, in 1857, the last remaining pocket of resistance to French rule in the mountains of the Kabylie was eliminated. The year 1857 is often considered the date that marks the total pacification of Algeria, although smaller rebellions continued to occur.[12] In 1858, under pressure from the settlers and the colonialist lobby, Napoléon III terminated military rule in Algeria and created a new Ministry of Algeria and the Colonies.[13] Over the next two years, Algeria was governed from Paris by a resident minister rather than by a governor-general in Algeria. During this short period, restrictions on property transactions were lifted to the benefit of the settlers, the Muslim population was denied access to Muslim justice and forced to deal with the French justice system, and efforts were made to assimilate the Algerian population in order to, in the words of historian Charles-Robert Ageron, "cut down the native aristocracy, weaken the authority of the *caïds,* and 'take the tribe to pieces.'"[14] The resulting disruption of the local economy imposed yet more hardships on the Algerian population, which felt increasingly alienated, frustrated, and threatened by French rule.

A relatively small group of men, the indigènophiles, many of whom worked in the Bureaux Arabes (Arab Bureaus), were gravely concerned by these disruptions of Algerian society and economy, and they notified Napoléon III.[15] In September 1860, Napoléon III visited Algeria himself to assess the situation. Based on what he learned during his short visit as well as from different informants (many of whom worked in the Arab Bureaus), he concluded that the settlers were greedy and had succeeded too well in convincing the civil colonial administration to facilitate their goals. Napoléon was concerned about the expropriation of land and resources such as forests from the Algerians and about their general impoverishment.[16] By November 1860 he had abolished the Paris-based Ministry of Algeria and in December he reestablished the military regime. A new governor-general, Marshal Amable Pelissier, duc de Malakof, was appointed and ruled Algeria until his death in 1864. Marshal Patrice MacMahon followed Pelissier as governor-general, ruling Algeria from 1864 until 1870.

In an effort to protect the Algerians from what he was convinced was expropriation and exploitation, Napoléon III put an end to the process of cantonment and worked to implement new policies that would protect the property and livelihoods of the Algerians. Laws were passed in 1858 and 1860, for example, that terminated the policy of providing free land to newly arrived settlers.[17] In 1863 he articulated his beliefs about Algeria by writing to Governor-General Pelissier that "Algeria is not, strictly speaking, a colony but an Arab kingdom," and that "the natives like the *colons* have an equal right to my protection."[18] So began the period of administration often referred to as Napoléon's Arab Kingdom. During this period, two important laws were passed under Napoléon's direction, laws that sought to protect both the property of Algerians as well as their legal status as persons. The first was the Sénatus-consulte of 1863, which decreed "the tribes of Algeria owners of the lands which traditionally they have always enjoyed."[19] The second law, promulgated after the emperor's second trip to Algeria in 1865, was the Sénatus-consulte of 1865, which "established the Muslims of Algeria as French, the equals of those with their roots in metropolitan France."[20] In addition, free land concessions for corporations were abolished in 1864 and limits were placed on the expansion of rural colonization.

Not surprisingly, the settlers and many in the civil administration were very much opposed to the policies of the "Emperor of the Arabs." Although they became disillusioned during this period, the settlers also fought the changes where they could. They manipulated details of the new laws, especially the Sénatus-consulte of 1863, to protect their interests in land acquisition. The settlers also had powerful contacts in Paris who fought for settler rights in Algeria. As a result of these efforts, by the spring of 1870 the French parliament had voted in favor of the settlers on a variety of issues.[21] That summer, however, France became embroiled in the Franco-Prussian War, and in September the Second Empire collapsed after Napoléon III's capitulation. In Algeria the resulting confusion, fueled by deep-seated discontent among most of the Algerian population, led to the Algerian Insurrection, which began early in 1871.

Many in the Arab Bureaus had been warning for years that the misery and frustration created among the Algerians by colonial administration would likely lead to revolution. These men were trained in the local languages and many spoke Arabic and the Berber dialects fluently. They tended to live closer to the Algerians, especially in the rural areas, than did the majority of settlers and thus frequently came to have a better understanding of and appreciation for their ways of life. Such specialists had been working for the military administration in Algeria since the early years of occupation. Beginning with the creation of the Arab Office in 1833 under the direction of Juchault de Lamoricière, the Arab Bureaus experienced several transformations and reorganizations.[22] In 1844 the Arab Affairs Office was established in Algiers, with an Arab Bureau staffed with military officers in each subdivision of every province in Algeria.[23] The Arab Bureaus were directed by a number of important figures in Algerian history, such as Pellissier de Reynaud (considered the "father" of the Arab Bureaus) and Eugène Daumas. Because they were perceived to side with the local Algerians, the Arab Bureaus and their officers attracted the enmity of the colonists from their inception.

Although many, and perhaps most, of the men in the Arab Bureaus put indigenous interests first, some did not. A few of these men used their positions for personal gain and some were documented to have physically harmed Algerians. Such incidents were used by the colonists and the settler lobby in Paris to try to curtail the power of the Arab Bureaus.[24] These incidents also generated a certain amount of resentment and enmity among the Algerians themselves toward the Bureaus in several parts of the country. The power of the Arab Bureaus was curtailed significantly after the 1871 insurrection and the subsequent "triumph of the settlers."[25] It is important to recognize, however, that for roughly forty years, from the early 1830s to 1870, the men working in the Arab Bureaus had a significant impact on the daily lives of rural Algerians—and on the landscape—via their policies and their administration of these territories. Many authors have concluded that the Arab Bureaus were the most powerful institutions shaping rural Algerian society during this period.[26]

Despite the fact that the settlers perceived the Arab Bureaus as defending the Algerians and working against European interests, the men working in the bureaus had their own specific views of what helping the Algerians entailed. They rarely defended traditional (precolonial) Algerian ways of living. Rather, the Arab Bureau officers shared with the colonists a vision of "proper husbandry" that included sedentary farming as the means to economic (market-integrated) progress. Thus, while they fought cantonment, for example, they advocated and worked toward the sedentarization of the nomads. Their goal was to establish villages where the nomads would settle and learn proper farming and livestock-raising techniques. They believed that this would "create a firmly-rooted peasantry of smallholders who would owe the security of their titles to France."[27] This would in turn facilitate the military's goal of security and pacification. Most Arab Bureau officers also shared with many of the settlers and other administrators the declensionist colonial environmental narrative that had begun to take shape during the first two decades of occupation.

The Colonial Environmental Narrative Refined

In 1848, at the beginning of the Second Republic, as noted earlier, the most common environmental narrative regarding Algeria and North Africa, was one that assumed both an innate fertility of the land and its waste by the Algerians and their "primitive" agricultural methods. Although a distinctly colonial environmental narrative had begun to take shape during the first two decades of occupation, it was not yet widespread, and the subsequent two decades to 1870 were pivotal to the refinement of what was to become the dominant environmental narrative of North Africa during the colonial period. This degradation narrative, rich with details of how the Algerians had deforested and overgrazed the landscape, was used by different colonial actors to rationalize and facilitate many of their goals and actions. Although most of the indigènophiles, including many in the Arab Bureaus, did try to understand and to support the Algerians, and worked to protect their rights and livelihoods, some of these men also played important roles in developing and spreading the declensionist environmental narrative.

The influence of the indigènophiles had been strong in Algeria from the early days of occupation. Many of the members of the Scientific Commission, for example, were indigènophiles. Several of these same men were also Saint-Simonians, including Prosper Enfantin, Ernest Carette, and J.-A.-N. Périer. These men promoted the utopian philosophy of Claude Henri de Rouvroy, comte de Saint-Simon (1760–1825). They believed that if the advances of science and technology were applied to Algeria, "the natural fertility and potential productivity of the region—so sadly unexploited hitherto—would once again flourish" with "the creation of a dense network of small, nucleated rural settlements."[28] Other indigènophiles, while not Saint-Simonians, shared many of these views on the "proper" kind of settled farming that would be best for the Algerians. A large number of these men were army officers, and several served in the Arab Bureaus, including Pellissier de Reynaud (also a member of the Scientific Commission) and Ferdinand Lapasset. As military men, it is not surprising that their primary goals included peace and security, predicated on pacification and thorough surveillance. This was much easier with settled populations. Captain Ferdinand Lapasset, for example, head of the Arab Bureau at Ténès, concluded his detailed plan for the organization of the Algerians in the military territories by explaining that administration by the Arab Bureaus provided "a greater security in the country; a more immediate surveillance; . . . greater submission of the Algerians; [and] all ideas of rebellion made impossible."[29] Lapasset was, in his own way, a champion of protecting the rights and property of the Algerians, and he fought cantonment. He also, however, preferred to sedentarize the nomads. Toward this end he directed the construction in 1845 of a model village the

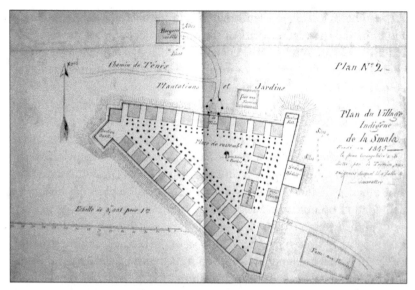

Figure 3.1. Plan for a nomad village *(plan du village indigène de la Smala)*. This village was built to sedentarize local nomads. It was conceived and constructed in 1845 by Ferdinand Lapasset, director of the Arab Bureau at Ténès. From Centre des Archives d'Outre-Mer (CAOM), ALG, GGA, 8H/10. Reproduced with the kind permission of CAOM.

purpose of which was the "progressive fixation of the natives to the soil."[30] Lapasset especially influenced Napoléon III's thinking and actions toward Algeria during the period of the Arab Kingdom (1860–70), as did the Saint-Simonian Ismail Urbain.[31]

The conventional narrative of untapped fertility was evident in most of the writings of the indigènophiles during the first several decades of occupation. Prosper Enfantin, one of the most important of the Saint-Simonian indigènophiles, wrote in the 1840s of the fertile soil and excellent climate of Algeria that would facilitate his vision of colonization.[32] Carette and Pellissier de Reynaud also shared the conventional view of a fertile environment that had not been properly husbanded and thus not reached its potential productivity. Some indigènophiles displayed a more sophisticated understanding of the traditional methods of agriculture and stock raising among the Algerians. While considering traditional property relations, for instance, Enfantin reasoned that one must take into consideration that the Algerians had been "pushed naturally, by the necessities of the soil and the climate, to the nomadic or pastoral life, and . . . to the sowing of cereals on land that is not consistently habitable."[33] For Enfantin and the vast majority of the other indigènophiles, however, nomadism was something to be curtailed, as the goals of colonization included the establishment of rational, sedentary, market-integrated agriculture and livestock production.

Similar sentiments against nomadic peoples were evident during the first two decades of occupation, as detailed above. Although not well described or widespread at that time, blame for overgrazing, deforestation, and desertification was assigned to nomads and their livestock by those in the military, some on the Scientific Commission, agricultural experts, and others. A few of these early accusations asserted a long history of degradation by nomadic pastoralists stretching back to the "Arab invasion" of the seventh century.[34] Well into the 1840s, though, the "Vandal invasion" of the fifth century received most of the blame for the destruction of North Africa in French writing about the region.[35] Related to this history of the Vandals was a widespread admiration for Arab/Muslim civilization whose establishment "in the west [of North Africa] had restored a new vigor to the civilization that had died, suffocated in the forceful grip of German and Scandinavian barbarity [the Vandals]."[36]

In 1853, however, after the publication of Carette's magisterial volume of the *Exploration scientifique de l'Algérie* on the origins of the principal tribes of North Africa, blame for the decline of North African civilization and the ruin of the environment began to be placed squarely on the Arab invasion rather than the Vandals. This publication was one of the earliest to make a very clear distinction between the seventh-century Arab invasion and the eleventh-century Arab invasion of the Beni Hillal and others from the east.[37] Drawing primarily on medieval Arab historians and geographers such as El Bekri, Idrissi, Ibn Khaldoun, and Leo the African, Carette recounted that "in 1048, three great Arab tribes marched in the number of one million persons. . . . Having arrived in the Maghreb . . . these nomads committed all sorts of abuses [and] looted to enrich themselves. . . . They drove away the inhabitants and ravaged the whole country. . . . For century after century it is easy to follow and to verify the progress of this inundation which upset everything, devastated everything, ruined everything. Quite different from the first invasion."[38] Importantly, Carette described environmental destruction by the invading "Arab nomads," proclaiming that "trees, cultivation, population, all had disappeared."[39] In other passages he described the destruction of plantations, the depletion of water sources, and the burning of trees. In contrast to what he had de-

Figure 3.2. The "Arab invasion." This illustration, from a chapter on "Arab domina-tion" in a mid-nineteenth-century history of Algeria, shows a seemingly endless stream of Arabs conquering North Africa. From Léon Galibert, *L'Algérie ancienne et moderne* (Paris: Furne et Cie, 1844), 125.

scribed as a relatively nondestructive first invasion, Carette stated that the "irruption of the eleventh century resembled a fire which . . . reduced everything to cinders, buildings and trees."[40] Carette quoted the twelfth-century Arab geographer Idrissi for the vast majority of such claims.[41] This is notable because later in the colonial period it was a different medieval Arab author, the historian Ibn Khaldoun, who was most commonly quoted by French writers in regard to the eleventh-century "Arab inva-sion" and its environmental destruction. Although Carette cited Ibn Khal-doun frequently, it was not for his descriptions of destruction but rather for genealogy and classification of the various Berber and Arab "tribes" of North Africa.

The writings of Ibn Khaldoun were not widely available in French until the Baron William MacGuckin de Slane, a translator for the French army in Algeria, published his translation of the *History of the Berbers and the Muslim Dynasties of North Africa* in four volumes from 1852 to 1856.[42] These volumes were only part of Ibn Khaldoun's corpus of fourteenth-century works known collectively as the *Kitab al-'Ibar* (*Uni-versal History*).[43] A section of the *Universal History* covering an earlier period, the *History of Africa under the Aghlabite Dynasty,* had been trans-lated into French before this by Adolphe-Noël Desvergers and published in 1841.[44] De Slane's first volume (1852) provided several descriptions of the destructiveness of the invading eleventh-century Arab nomads that became widely quoted.

Most of these passages described the activities of two groups in particular, the Hillalian (Beni Hillal) and the Soleim. According to de Slane's translation, Ibn Khaldoun wrote that the "Hillalian families swooped down on Africa like a swarm of locusts, spoiling and destroying all they found in their passage."[45] They "devastated the gardens and cut down all the trees" in many locations.[46] "This territory is now changed into desert. . . . Before, the Sanhadja dynasty had made agriculture prosper, but the Arab nomadic pastoralists brought devastation and succeeded in constricting, by their incursions and their plunder, the limits of the cultivated area."[47] Most of these quotes, however, referred to quite specific locations, very often in Libya, although they were generalized to all of North Africa by later writers. These passages, and others were used throughout the colonial period—some even to the present day—to illustrate the colonial story that the invading Arab nomads had ruined North Africa. As did Idrissi before him, though, Ibn Khaldoun also praised nomads and the Arabs in general. In fact, elsewhere in his *Universal History* he states that civilizations are born of nomadic societies and that nomads are more "good" and courageous than sedentary peoples.[48]

Considered in their entirety, the writings of these medieval Arab authors are subtle and complex, and they must be interpreted within their historical, economic, and political contexts. Postcolonial critics have revealed the selective use by colonial authorities of indigenous writers in many parts of the world for their own purposes.[49] French geographer Yves Lacoste, for example, went to great lengths to document and refute the colonial construction of what he termed "the myth of the Arab invasion" in his book on Ibn Khaldoun.[50]

The eleventh-century Arab invasion, also called the Hilallian "catastrophe," has been frequently cited by French authors and called the "most important event of the entire medieval period in the Maghreb."[51] According to one author, the invasion "marked, without question, the start of a new era"—after which, due to the destruction, nothing was the same again.[52] In the words of geographer Émile Gautier, "the immigration of [Hillalian] nomads was the immense catastrophe, the end of the world."[53] In the majority of these and other accounts, the authors cite Ibn Khaldoun for proof of the wholesale destruction wrought by the Arab invaders. Ibn Khaldoun does not, in fact, say anything of the kind, and Lacoste demonstrates how various writers selectively chose parts of Ibn Khaldoun's writings to support their goals while deliberately omitting other sections.

Lacoste, writing in 1966, lamented the fact that "despite the obvious complexity of the issues involved, the vast majority of historians of North Africa still subscribes to the thesis that the 'Arab invasion of the eleventh century' destroyed the achievements of the sedentary population."[54] He explained that "turning the Arabs into invaders was one way of legitimizing the 'French presence'" and that it "provided a historical basis for a policy of turning Arabs and Berbers against each other."[55] Historian Patricia Lorcin has made the related argument that, during the colonial period in Algeria, an influential myth emerged that the "Kabyles were superior to Arabs" and that "the French used sociological differences and religious disparities between the two groups to create an image of the Kabyle which was good and one of the Arab which was bad."[56] She and other authors have termed this the Kabyle myth.[57]

The importance of the environmental narrative inherent in both the myth of the Arab invasion and the Kabyle myth has not yet been widely recognized. Not only were the Arabs portrayed as ruining the eleventh-century Berber civilization, but they were also blamed for ruining a previously fertile and forested environment.[58] Since the Arabs allegedly had retained most of their wandering, anarchic, destructive habits, many argued that they also continued to damage the environment into the colonial period. Thus, in one theme of the colonial environmental narrative, the sedentary, agricultural Berbers (mostly the Kabyles) are portrayed as good, or at least better, stewards of the environment than the destructive, nomadic Arabs.[59]

This view of the Berbers is illustrated in popular orientalist artwork of the period, of which Eugène Fromentin's *Kabyle Herders* (*Bergers kabyles*) provides a representative example. Dating from the 1860s or 1870s, it depicts two herders in a relatively lush setting in the Kabylie, with many trees, shrubs, and other plants bordering a stream (see plate 3). Fromentin (1820–76) shows them resting, one playing a flute, while two sheep stand nearby, looking well-tended, fat, and calm. The colors are rich but cool, dominated by deep blues, greens, and golden browns. None of the trees or other plants shows any signs of overgrazing, and the overall effect of the work is one of a peaceful, happy, abundant mountain landscape. Fromentin's image represents the Berber herders as a contemporary perpetuation of the ancient pastoral tradition.[60]

Although the primary components of the colonial environmental narrative had been developed and published by the mid-1850s, the fully articulated declensionist narrative was not often used until later in the colonial period. Parts of it, though, were used by those who perceived a benefit from invoking them. As earlier in the colonial period, those who benefited most appear to have been the colonists, the colonialist lobby, government administrators, and some military men. Among government administrators and those commissioned to produce studies for the government during the Second Empire, a varying mix of the different parts of the colonial narrative can be found. For example, a study of the physical, economic, and political geography of Algeria commissioned by the War Ministry in 1849 displays many of the early parts of the narrative but not those developed in the 1850s. The author of this comprehensive study, Oscar Mac Carthy, extolled the fertility of the soil and wrote that the former "granary of Rome has lost little of its fertility."[61] Although he lamented the state of the forests, "devastated by the hand of man, ravaged by the teeth of beasts," he did not specify the Algerians or the nomads as the villains. The section detailing indigenous forest use does not describe any destructive practices by the Algerians.[62] Nor does the story of the destructive eleventh-century Arab invasion appear in this volume.

That story does appear, however, in a work by Jules Duval, former magistrate and secretary to the General Council of Oran Province. In his own comprehensive overview of Algeria, Duval described the "invasion of the Arab hordes," the "Arab torrent [that] destroyed villages, ravaged the countryside, plunged into a most profound misery and barbarism a country that still bore numerous traces of previous civilizations."[63] Elsewhere in his book, however, he invoked the more positive parts of the narrative. Duval characterized Algeria as the granary of Rome, praised the fertility of the soil, described the vegetation and forests as magnificent, and portrayed the Sahara as fertile in pastures rather than as empty, degraded, or desertified. Duval also painted a positive, rather romantic image of contemporary Arabs directly contrary to the Kabyle myth: "[T]raveler, herder, nomad. . . . His spirit has remained more elevated than that of the Kabyle, his imagination more lively."[64] Duval portrayed Algeria in this largely positive way because his purpose was to make Algeria better understood by those in France who, he believed, had a misconception of it. He specifically wanted to encourage immigration from France to Algeria rather than to North America, for example.[65] Mac Carthy, too, strongly favored colonialism and immigration, and these positions also likely motivated his positive descriptions and inclusion of the positive parts of the environmental narrative.

Ismail Urbain, perhaps the most influential of the indigènophiles, and a Saint Simonian, was persuaded by several of the more negative parts of the colonial environmental narrative. It was Urbain who provided a majority of the ideas that formed the foundations of Napoléon III's pro-Arab policies; he also held an influential position as a member of the governor-general's Advisory Council.[66] Although a steadfast and outspoken champion of Arabs and Islam, like most of the other indigènophiles he believed that settled agriculture and stock raising were superior to extensive pastoral nomadism. He not only incorporated a description of the destructive eleventh-century Arab invasion in his writing but also criticized the way of life of contemporary Arab nomads.[67] He clarified his position later in the book when he explained that cantonment was indispensable because agricultural conditions had improved under the French as the nomads had settled down, made gardens, and become "less nomadic, their interests fixed on the soil."[68] Therefore, he concluded, they no longer needed large amounts of land. Despite his reputation as an *Arabophile,* his support of the policy of cantonment was counter to the views of several other prominent Arabophiles, such as Lapasset.

While some colonial actors focused on the environment as a whole, including agricultural fields, pastures, and the like, others tended to focus solely on forests and areas thought to be deforested. As might be expected, the colonial narrative was often invoked in their writings. The French inspector of colonization in Algeria, Céleste Duval, lamented in 1859 that "if the Arabs, for centuries . . . instead of destroying, burning, cutting, had imitated the descendants of the ancient Talestris, . . . Algeria [and] the northern coasts of Africa, would not have to deplore their bareness and their solitude; they would be adorned and revived with these noble objects of affection [trees] of our ancestors and our brothers."[69] While not explicit in dating his accusations, Duval claimed that the Arabs had been deforesting North Africa for centuries, probably dating back to the Arab invasion. Duval also blamed many of the colonists themselves for deforesting. His primary concern was with reforestation as the best way to restore the fertility of Algeria and to improve the climate and overall healthfulness and productivity of the colony. He believed that in the past the coasts of North Africa had been heavily forested and "resplendent with the magnificence of nature."[70] He noted, though, that the Forest Service had been badly treated, which had prevented it from protecting or improving Algeria's forests.[71]

Little was published on Algerian forestry during the period of the Second Empire. The forests were not yet well explored or inventoried and access was still difficult and dangerous in many areas. The only significant work appeared as a secondary part of a major volume on French forestry, published by a professor in the National Forestry School at Nancy in 1860. Primarily a botanical treatise, it contained little commentary and no mention of any part of the environmental narrative of North Africa.[72] Another work, the most important of its time on Algerian botany, is notable for its nearly complete lack of the colonial environmental narrative despite its extensive commentary. Published in 1867 by two well-known and respected French botanists, it was part of the *Exploration scientifique de l'Algérie* and one of the last volumes published in the series. Although the bulk of the monograph is devoted to the typical botanical descriptions and classifications, the seven-page "Note sur la Division de l'Algérie en Régions Naturelles" provides a rich commentary on the Algerian environment.[73] Here the authors described the "beautiful forests" found in many of the mountain regions and the "fertile plains" found in others. In regard to areas such as the southern High Plateaus, often described later in the colonial period as deforested or otherwise ruined, these authors wrote that the scarcity of vegetation was determined by nothing other than the high-altitude conditions that characterize the region.[74]

In their discussion of the Sahara Desert and its vegetation, they carefully pointed out the natural limits of the desert by latitude and longitude, adding that "the limits of the Algerian Sahara are more political than natural," and that "the fertility of the soil is determined almost exclusively by the presence of water."[75] They believed that the distribution and type of vegetation in the Sahara confirmed "the law after which the influences of latitude are dominant in the interior of Algeria."[76] There was no mention of Arabs or Kabyles or the eleventh-century invasion, and no discussion of the ruin of the Algerian environment at the hands of humans. Only a brief reference is made early in this note to the fertility of the soil and the thickness of the forests. Although part of the Scientific Commission, neither of these botanists lived in Algeria; they visited the country only to make botanical studies. Perhaps, then, they did not have the vested interest in colonization efforts that seems to have led others to use the colonial environmental narrative. There was little else published on botany during this period, likely due to the fact that no professional botanists resided in Algeria until the mid-1870s.

Resident immigrants, on the other hand, provide some of the most vitriolic examples of the utilization of the colonial environmental narrative during the period of the Second Empire. Many of these men were heavily invested in the successful colonization of Algeria and felt they had a lot to lose if too many "concessions" were made to the Algerians, especially during the decade of Napoléon III's Arab Kingdom. One influential colonist, François Trottier, an avid arboriculturist in the Mitidja, wrote several books and pamphlets on deforestation, afforestation, and colonization, particularly on eucalyptus, during the 1860s and 1870s. Primarily concerned with the deforestation of Algeria and the urgent need to reforest it (thus the need for his Eucalyptus plantations), he cited the assertion by the French agronomist de La Tréhonnais that "in the past . . . North Africa was clothed with dense forests."[77] He explained that deforestation was caused primarily by the nomadic pastoralists because "the forest is noxious to the herder."[78] He added that the Arabs burned the forest not only to create pasture but also out of malice and that they cut down trees to use as tent stakes. The only major part of the colonial narrative missing from this account is the eleventh-century "Arab invasion," although the multitude of disparaging remarks throughout his writings about the Arabs certainly implied that they had been destroying the Algerian forests for centuries. The solutions he proffered to remedy the situation focused on outlawing nearly all indigenous uses of the forests (especially grazing, burning, and collecting bark for tannin production), eliminating communal property, and, by implication, destroying the traditional Algerian way of life. After all, Trottier concluded, "for land that is fertile or covered with trees . . . the Arab is a plague, he has always been so and he will be in the future; civilization must annihilate him, because he exists against providential destiny."[79] Despite his rhetoric extolling the benefits of forests generally for climate and civilization, Trottier's primary motivation was economic, driven by his investment in eucalyptus.[80]

On the eve of the Franco-Prussian War, in December 1869, a short book was published by another colonist, Henri Verne, who was a corresponding member of the General Council of Constantine. This booklet is notable for its statement of many of the most important complaints and demands of the settlers unhappy with Napoléon III's Algerian policies. It is also significant because it contains one of the most complete iterations of the declensionist colonial environmental narrative to be found during the Second Empire. Explaining why the Arabs needed to be immobilized and civilized, Verne proclaimed that when the Arabs, "the Muslim hordes," invaded North Africa, "fire destroyed the harvests, the plantations, the forests, and a society newly established on this land was laid to waste."[81] He described the fertile soil, mentioned that this region had been the granary of Rome, praised the indigenous Berbers for their tree-loving habits, and condemned the Arabs for burning fields for fertilizer and forests out of malice. Following his description of the destructive Arab invasion, Verne explained pointedly that, due to their many vices (Islamism, communism, polygamy, fatalism), the Arab race is always decreasing. He concluded that the Arabs had "arrived at the point of collapse where [they do] not even know how to plant a tree."[82] Verne used every major part of the colonial environmental narrative in this short book, in part to assert that it was incumbent upon the French to lift the Arabs out of "this infantile state" and "to attach them to the soil by the possession of property."[83] He then argued for the benefits of cantonment, against the detrimental effects of the Sénatus-consulte, and promoted other aspects of the settlers' agenda.

By the late 1860s, then, the major components of the declensionist colonial environmental narrative had been developed and were being invoked by various actors for different purposes. Although few put the entire narrative together at once during the Second Empire, many did so following the Algerian Insurrection of 1871. The earliest to use the narrative in its most complete form appear to have been the settlers and the colonialist lobby. They were also often the most vicious in their wielding of the story. As the Second Empire gave way to what would become the Third Republic, the colonial narrative became progressively more entrenched, widespread, and in many instances tightly related to the Kabyle myth. The idea of the Arabs as nomadic invaders who wreaked environmental destruction, combined with the idea of the Berbers as the original benevolent, sedentary inhabitants, would lead many to the "rational" conclusion that the Arabs must be immobilized and civilized by any means possible. As a result, the French pacified and attempted to sedentarize nomadic populations not only in Algeria but also later in Tunisia, Morocco, and many of their other colonial possessions.[84] This prejudice against nomads was closely related to similarly important biases against fire in France, Europe, and North America in the nineteenth century. Such attitudes were in turn related in complex ways to the forestcentric mood that gripped much of the Western international scientific community and lay people alike, as it still does today.[85]

Emerging Environmentalism and the Colonial Narrative

Nineteenth-century international scientific and learned ideas about the environment, and particularly those about forests, bear directly on the French colonial environmental narrative of North Africa. In France, much of Europe, and the Americas, concern for protecting nature for environmental reasons was on the rise and competing with earlier, primarily utilitarian reasons for nature conservation, in some instances replacing them. Although the origins of environmentalism have been traced to the eighteenth century, it was during the nineteenth century that these ideas started to become widespread.[86] Deforestation preoccupied many scientists and politicians and "desertification" was closely associated with deforestation. It was common for people to believe that deserts were unnatural aberrations that should and could be forested.

By 1864, when George Perkins Marsh published his widely influential *Man and Nature*, concern about deforestation and the conservation of forests had become ubiquitous in Europe and North America. Indeed, a forestcentric mood gripped much of the international scientific and elite community. This attitude is illustrated by Marsh's proclamation that "there is good reason to believe that the surface of the habitable earth, in all the climates and regions which have been the abodes of dense and civilized populations, was, with few exceptions, already covered with a forest growth when it first became the home of man."[87] The sources Marsh cited for this claim are both French authorities: Louis-Ferdinand Alfred Maury and Antoine César Becquerel.[88] The significance of these ideas for North Africa became clear a few pages later, when Marsh stated unequivocally, "I am convinced that forests would soon cover many parts of the Arabian and African deserts, if man and domestic animals, especially the goat and the camel, were banished from them."[89] Marsh relied heavily on French sources, but also cited other prominent European authors of the period, and his inclusion of these English, German, Scandinavian, and American sources makes Marsh's book a very useful synthesis and reflection of international thinking about the environment in the middle of the nineteenth century. Marsh's book was quite influential in the international community as well as in the United States.[90] It was widely quoted by French

and other European writers, particularly when they discussed the importance of forests. The French geographer Elisée Reclus, for example, was
influenced by Marsh's work, and his *The Earth and Its Inhabitants* (1887)
was cited often as providing evidence of the environmental deterioration
of North Africa.[91] Marsh was also quoted with great enthusiasm by the
Algerian colonist and ardent reforester Trottier, who was of the opinion
that Marsh "has made the most complete investigation into the disasters
caused around the globe by the destruction of forests."[92]

The "tree-centric" attitude evidenced in *Man and Nature*, then, was
widespread throughout Europe and North America during the nineteenth
century. Forests had not always, however, been the focus of affection and
conservation. In fact, attitudes toward trees and forests in the Western
world had changed significantly in the preceding two millennia.[93] Serious
attention was being given to trees and forests as something valuable and
in need of protection by the fourteenth century, because of the growth of
shipbuilding and ocean-going naval power. In the seventeenth century
France was acting at the state level in this regard: the Forest Ordinance of
1669 restricted many traditional uses of the forest in order to preserve the
woods for timber production.[94] Five years earlier, in 1664, the Englishman
John Evelyn wrote his influential *Silva: Or, A Discourse of Forest-Trees*,
both an historical exposé of the degraded state of English forests and a
how-to manual for their conservation.[95] Although there is earlier evidence
of concern for conserving particular forests, for example in the twelfth
and thirteenth centuries in France, there were also significant negative attitudes toward forests in the Middle Ages and earlier.[96] Conflicts frequently
occurred between peasant farmers who needed land cleared of trees for
agricultural production and elites who preferred to retain and enlarge
forests for game preservation and hunting. The Magna Carta, for example, drawn up in 1215, contained an article requiring that some royally
afforested areas in England be cleared and returned to their "original,"
usually agricultural, condition.[97] Before this period and after, forests were
also often seen as dangerous places to be cleared or avoided, either for religious reasons or because villains and thieves frequented them. Clarence
Glacken dates the change in attitude, from one that preferred forest clear-
ance to one that preferred forest conservation, to around the thirteenth
century.[98] Attention to the interaction between forests and the environment, though, dates back to the ancient world. During the classical period,
for example, Theophrastus worried that clearing forests had changed the
climate, and Plato seems to have bemoaned apparent deforestation and

significant erosion in Attica.[99] The history of Western thinking about forests, then, is long and complex, and the ubiquitous forestcentric attitude of the nineteenth century was something relatively new.

Notions regarding forests and their conservation and use were also tightly bound up with perceptions of fire and pastoralism, both of which were widely believed to be harmful to forests. Antifire attitudes and practices in Europe have been linked to the rise of forestry legislation in the seventeenth century, particularly in France. Stephen Pyne has written that "forestry had demonized fire."[100] While the French Forest Ordinance of 1669, and later the 1827 Forest Code, were historic legal developments that placed many formal prohibitions on the use of fire in and near forests, earlier antifire attitudes and practices have been documented. In fifteenth-century France, for example, lighting a fire in the forest of Orléans was punishable with a hefty fine.[101] Indeed, it appears from the historical record that France was one of the first countries to adopt antifire policies in order to protect economically valuable forest resources. Pyne argues that France "pioneered fire protection."[102] Forest fire regulations and prohibitions were present in the 1669 and 1827 legislation and also in separate laws enacted during the late nineteenth and early twentieth centuries, specifically the French laws of 1870 and 1924 and the Algerian law of 1874.

Throughout European history, fire had been used quite commonly to prepare agricultural fields for planting as well as to revitalize pastures. The use of fire was closely associated with transhumant livestock raising, especially in the Mediterranean regions of Europe, where forested areas were not infrequently burned to create pastures.[103] This association between pastoralism and fire has a long history dating back to at least the classical period. Due to the long history of association between pastoralism and fire in Europe, many held negative attitudes toward pastoralism, especially transhumant pastoralism.

Negative attitudes toward transhumant and nomadic pastoralists in France and Europe may thus be traced, at least partly, to antifire attitudes common in the nineteenth century. However, antinomad attitudes in Europe were also motivated by other ideas with long histories. One of the most powerful and widespread was that of the stages of civilization.[104] Societies and peoples were widely thought to progress from the "primitive nomadic" state to the "civilized sedentary" state. Auguste Comte, for example, wrote in the nineteenth century that "the prime human revolution is the passage from nomadic life to the sedentary state."[105] Similar ideas may be found in the writings of several nineteenth-century European authors, including Hegel, Darwin, and others. Many of these authors, especially Comte, were influenced by the Marquis de Condorcet, who in the eighteenth century devised an elaborate outline of the progress of humankind—from the lowest stage, of barbaric hunters and gatherers, on through stages including pastoralism and agriculture, to the ninth stage of scientific reason and civilization epitomized by postrevolutionary France.[106] In this formulation, not only were nomadic pastoralists at a more primitive social stage, but their mode of production (mobile livestock raising) was also primitive and irrational compared to the more advanced stage of settled, scientific agriculture.

Negative ideas of nomadism may be traced as well to long-standing antagonisms between mobile and settled peoples dating back to antiquity. Wandering people and those closely associated with livestock were frequently perceived to be outside the limits of civilization, irrational, and therefore a threat to the sedentary social order.[107] This attitude toward nomads was still very much evident when Herbert Spencer wrote, in the nineteenth century, that those who live "by [the] kepinge of cattel . . . are both very barbarous and uncivill, and greatly given to warr [*sic*]."[108] Nomadic pastoralists were ranked quite low in Spencer's influential theory of the evolution of human societies toward complex, sedentary civilizations. Also associated with these ideas were ethical judgments that nomads were immoral because they chose to live outside the social order, that is, because they preferred the uncivilized life. Thus, in the words of John Noyes, "since it was widely felt that European civilization emerged through a process of sedentarization, moralization and rationalization, nomadism came to be seen as both a historical and a geographical (or territorial) limit to civilization."[109] In the context of nineteenth-century Europe and the expansion of Western capitalism, nomadic production systems were also perceived as especially irrational and inefficient since they were, supposedly, not based on a system of private property.

For the French in Algeria, many of these ideas coalesced into the general sentiment that nomads were barbaric, backwards, irrational, dangerous to society, harmful to the environment, and in need of civilizing. The process of civilizing them had to include the sedentarization of the nomads as well as their incorporation into the international market economy, including the establishment of private property and rational farming methods. The colonist Henri Verne captured this sentiment well.

> Thus will civilization advance. . . . It follows that the establishment of private property and the extension of colonization are indispensable where the Arabs are concerned; that military authority must maintain the native in a state of immobility which will destine them to extinction. A barbaric people cannot find themselves faced with an advanced civilization without engaging in a duel to the death. Civilization must conquer barbarism or perish itself. Let us work therefore to transform this race: there lies our interest, there also lies their salvation.[110]

French colonial thinking about environmental history also needs to be considered in the context of some common mid-nineteenth-century religious and moral ideas about the environment. As Richard Grove has argued persuasively, the European image of the tropical island with its lush vegetation came to represent an ideal landscape with very positive moral connotations in the eighteenth and nineteenth centuries.[111] The forested tropical island was for many at the time closely associated with the Garden of Eden. Furthermore, in the early and mid-nineteenth-century European thought was still dominated by the conception of biblical time. As calculated in the mid-seventeenth century by Archbishop Usher, the world was believed by many to have been created in 4004 BCE.[112] From that point, when most of the habitable world was thought to have been forested, it was widely presumed that the world had suffered a long, slow decline, a fall from its "forested golden age."

Just as forests were associated with "good" and the image of the Garden of Eden, treeless landscapes, especially deserts, were associated with "evil," or at least with immorality. As Grove put it, there were even those who "conceived of drought as the wages of environmental sin or sins of moral disorder" and who saw "drought as a form of moral retribution" against the indigenous population.[113] In France, for instance, François-René, vicomte de Chateaubriand, the famous writer who became one of Napoléon Bonaparte's ministers, explained that, "everywhere the trees have disappeared, man has been punished for his improvidence. I can tell you better than others, sirs, what is produced by the presence or absence of forests, since I have seen . . . the deserts of old Arabia where nature appears to have breathed its last breath."[114] Thus, not only had landscapes without trees suffered directly at the hands of decadent and "immoral natives," but they were also literally signs of divine retribution. In the case of North Africa, it was the Arab nomads who were almost universally held responsible for creating deserts.

Although the word "desertification" would not be used until the early twentieth century, the idea of desertification as a process that began with deforestation and resulted in the complete loss of vegetation, and ultimately the degradation of the soil, was common in nineteenth-century French North Africa. Pastoralists and especially nomads were painted as the most common villains in this process. Officer Poinçot, for instance, referred to in the previous chapter, blamed the nomads for destroying the soil as early as 1834.[115] By midcentury, many in Algeria described large parts of the colony as a "land of thirst" (*pays de la soif* in French and *bled al ateuch* in Arabic). Auguste Warnier wrote in 1865 that the land of the nomads was "long ago a sort of terrestrial paradise. . . . Today this land is a sterile desert."[116] In 1880, the forester J. Reynard declared in his widely influential publication on the south of Algeria and the nomads there that "the native has created the dunes and the sand."[117] In fact, fears of desertification, specifically of the Sahara Desert expanding northward and threatening the prosperous agricultural areas of the Maghreb territories, were frequently articulated during the last half of the century. One impassioned advocate of reforestation, for example, warned that "the Sahara, this hearth of evil, stretches everyday its arms towards us; it will soon enclose us, suffocate us, annihilate us!"[118] The idea and the process of what would become known by 1927 as desertification thus has deep roots in the French experience in North Africa.

Not all of the French, though, saw the desert as a ruined landscape or as a place of danger and death. Long traditions of French art and literature present the desert as a site of beauty, divine inspiration, sanctuary, even as a proving ground of individual self-worth.[119] Many of the artists and writers responsible for such interpretations were and are still referred to as "orientalists." A number of French orientalist artists were drawn to the deserts of North Africa, especially after Algeria was pacified in the 1850s.[120] The subjects of their desert paintings ranged from landscapes empty of humans or animals, to nomad caravans, to hunting expeditions, to oases, to portraiture. Numerous French painters, many of them less well known, created paintings of the desert in Algeria that depicted exotically beautiful and even cheerful scenes. Adolphe-Pierre Leleux's 1853 *Threshing Wheat in Algeria* is a good example (see plate 4). Leleux (1812–91) traveled for six months in Constantine and environs in 1847. This painting, made with a grant from Napoléon III's government, captures the raw energy and happiness that accompanied many harvests in the desert.[121] Such harvests could occur only in the rare years of exceptional rainfall and were thus occasions to celebrate. This work and many others like it, however, were painted before the declensionist environmental narrative became ubiquitous in the 1860s and 1870s.

In 1861, Léon Belly painted *Pilgrims Going to Mecca,* one of the earliest "true desert" landscape scenes of North Africa that gained acclaim (see plate 5). Belly (1827–77) wrote that he was going to make a "painting representing a desert scene with Arabs and dromedaries. Penetrate rather deeply into study of the figure to reproduce what is truly beautiful and interesting in the present life of our humanity."[122] This painting is indeed beautiful. A caravan of pilgrims dominates the center of the painting, with a convoy of camels and people trailing far into the distance. Camels and pilgrims are painted in great detail, in strong colors, and are shown in robust good health and with noble bearing. The landscape, by contrast, is rendered in muted colors, with a sandy desert foreground rimmed by a purple haze of mountains in the far distance. A few small stones punctuate the landscape, and a single camel is nibbling at one of the few dried-out bits of twiggy vegetation found in the tableau as several large, indistinct birds circle overhead. There are hints, though, of the desert as a dangerous place. The light is bright and the sky is cloudless, hazy, and oppressive, creating a feeling of intense heat. The whitewashed bones of an animal of some kind are scattered along one side of the caravan. Given this evidence of death, the birds circling overhead could easily be interpreted as vultures waiting patiently for the next victim of the desert. One might also see a suggestion here that camels such as those in this caravan have "overgrazed" this landscape; thus the potential significance of the lone camel trying to find something to eat. This suggestion of a ruined landscape made into desert becomes much stronger and more explicit in later paintings.

Gustave Guillaumet's 1867 *Le Sahara,* considered a key painting in the orientalist genre, is one such work (see plate 6). Guillaumet (1840–87), one of the most famous French orientalists, painted this canvas when he was twenty-seven, after several trips to Algeria amounting to a considerable amount of time spent in the colony. It was very well received when it was exhibited in the Paris Salon. At the center foreground of Guillaumet's canvas appear the bones of a long-dead camel. Sparse vegetation dots the landscape, while a faintly visible group of people with camels (probably nomads) cluster in the distant background right of center. Painted in rich and beautiful browns, pinks, yellows, and blues, the work has generally been interpreted as illustrating the great emptiness, violence, and danger of the desert. It has also been interpreted, though, as a "more positive image of the desert as a spiritual haven of centuries-old tradition and as a source of divine inspiration, born out of the travails and challenges of life in the desert void."[123] Considered in the context of the prevailing wisdom of the time regarding the environmental history of North Africa, however, a third interpretation can be made. This tableau can be seen as illustrating the destructiveness of the nomads, seen here leaving an over-grazed area desertified by centuries of abuse. A camel and rider collapsed at the left of the canvas appear to have been callously left behind to die. The bones of the camel (the hardiest and toughest of the nomads' livestock) at the center of the painting may also suggest the coming end of the nomads' way of life; such an interpretation is further dramatized by setting the scene at sunset.[124] The imagery thus suggests that the destructiveness and brutality of the nomads' way of life must end. For those sympathetic to most colonists and administrators in Algeria, those who sought and actively worked to stop the nomads' wandering and ruinous ways, cultural imagery like this provided a kind of ideological support for many legislative and policy actions of the time. Indeed, many of the paintings Guillaumet exhibited in the salons of the 1860s were "positively received by the state as illustrative of its Algerian policies."[125]

Another painting that has been interpreted as offering support for the colonial project in Algeria is Eugène Fromentin's 1869 *The Land of Thirst* (*Au Pays de la soif*). Painted more than two decades after Fromentin made the first of several visits to Algeria and about fifteen years after he published an account of his travels in the Sahara, it is a bleak canvas.[126] The painting depicts a group of dark-skinned men who have perished in the desert (see plate 7). Five men and one horse, all dead or dying, fill the foreground. Several large birds fly overhead, seen against sparse clouds, arriving from the far distance. The landscape, a rough, stony desert with many small hills and valleys, contains not one trace of vegetation. Fromentin used bold colors in this canvas, for even the sandy terrain is a vibrant golden brown. The painting presumably took its name from the final sentence of his 1857 book, *A Summer in the Sahara:* "I will salute with a profound regret this menacing and desolate horizon that has been so precisely named—*Land of Thirst.*"[127] An engraving of this second version of his painting was chosen to illustrate an 1879 volume containing *A Summer in the Sahara.*[128] An earlier version of the painting has been interpreted as a "plea for the reconstituted imperial France to rescue the dying civilization of Algeria" and "was thus part of the discourse of French imperialism."[129] It certainly appears that both versions depicted a land and a people expiring and in need of help. It is also very likely that this painting consciously illustrated aspects of the colonial declensionist environmental narrative, making this a depiction of a land that has been so deforested and overgrazed that it can no longer support human or animal life. Fromentin was almost certainly familiar with the dominant narrative and likely believed it; his paintings provided imagery that perpetuated it, reinforcing arguments for its attendant imperial goals. In any case, Fromentin was quite critical of what he considered the Algerians' hatred for trees. He condemned the "passersby [who] lop them, the herders [who] mutilate them, the wood cutters [who] make firewood out of them."[130]

Both *Le Sahara* and *Au Pays de la soif* were highly influential. *Le Sahara* was very well received when it was shown in 1867 and became one of the cornerstones of French orientalism. Although *Au Pays de la soif* was not exhibited publicly until 1892, during a resurgence of the genre, an engraving of it was used as noted above to illustrate Fromentin's 1879 volume, *Sahara et Sahel.* This book was extremely popular and enjoyed multiple editions.[131] The engraving thus reached a very large audience even before the painting was shown in 1892. Both paintings remain popular to this day and have been subjected to a broad range of interpretations that see in them everything from positive images of spiritual havens, to desert wastelands, to parables.[132] Whatever the merits of those interpretations, both paintings illustrate the French colonial declensionist environmental narrative, consciously or not. In painting the desert as a ruined landscape, Guillaumet, Fromentin, and other orientalist artists helped to naturalize (primarily for the Parisian public) a story that was being actively used to bring about many changes in Algeria.

The nineteenth-century worldview, constrained by belief in biblical time and the fall from an earlier (forested) golden age, precluded most alternative conceptions of the environmental history of the Mediterranean basin, especially of North Africa. A much longer view of the history of the region, one that might entertain the concept of vegetation changing dynamically over time, was not possible for most Europeans before the end of the nineteenth century. Charles Lyell's *Principles of Geology* had challenged biblical conceptions of time as early as 1830, when its first volume appeared, but it was not widely influential until the late nineteenth century.[133] The new conception of geologic time (in which the earth was much older than 6,000 years) did not begin to gain popular acceptance until the 1870s and 1880s, when prehistory and geographic/geologic cycles began to be more widely discussed.[134] As the new thinking about geologic time gained hold, it precipitated a crisis of belief among many Christians—a crisis that was frequently articulated as fear of deforestation, climate change, and environmental collapse, and, thus, fear of the deterioration of humankind itself.[135] This was certainly the case in French Algeria. In the words of one apprehensive Algerian colonist, "the forested areas . . . must be tripled; it is by this condition that our race will conserve its European capacity; we must respond against the climate if we don't want to be degenerated and absorbed by it; the tree must be our anchor of salvation."[136] Salvation, for the Europeans, was increasingly believed to be found in the suppression of fire, the sedentarization of nomads, and the planting of trees.

Science and the Development of Forestry in France and Algeria

Fire and grazing suppression, as well as reforestation practices, have deep roots in French forest policies. French forest ordinances and forest legislation represent some of the earliest such laws in Europe. Economic and political goals provided the primary motivations for the two landmark pieces of forest legislation in French history, the 1669 Forest Ordinance and the 1827 Forest Code. Each focused on the protection of woods primarily for the production of naval timber and other valuable forest products. Similar regulation can be traced back to the early thirteenth century, most of it also motivated by economic and/or political concerns.[137]

Jean-Baptiste Colbert, finance minister under King Louis XIV, best articulated the urgency of the matter of forest conservation when he held that "France will perish for lack of wood." His primary concern was ships for the French navy. He alerted King Louis XIV, who convened a forestry reform council that worked from 1661 to 1669 to produce the 1669 Forest Ordinance, which is noteworthy for its comprehensiveness.[138] It had as its goal the protection and management of all forests in France, to a greater or lesser degree, including state (royal) forests and those belonging to the church, individuals, or communities (communal forests). Significantly, the Ordinance banned or restricted a great many traditional activities in and uses of the forests. The grazing of sheep and goats, for example, was banned in all forests in France, as were the burning of trees or kindling of fires.[139] Infractions of the Forest Ordinance were to be punished with steep fines, but enforcement was uneven for the next century; in some parts of France it was not enforced at all.[140] The Forest Service was reorganized so as to be based on a system of forest masters (*maîtres des eaux et forêts*), forest officers, and forest guards that dated back to the thirteenth century but that, until the reorganization, was not universal.[141] The Forest Service was changed dramatically in the wake of the Revolution, however. A law passed in 1791 eradicated the position of forest master and created a new, more centralized Forest Administration.[142] Still, despite some additional changes brought by the postrevolutionary government (for example, increased protection of communal property), the 1669 Forest Ordinance remained the official law, as it was never abrogated. It thus formed the basis for the reforms that resulted in the 1827 Forest Code.

Napoléon Bonaparte also viewed the forests primarily as economic resources, and housed the new Forest Administration that he created in 1801 under the Ministry of Finance. As with his predecessors, his primary concern was naval shipbuilding. He extracted state revenues from the forests as well, and used the 1669 Forest Ordinance to protect both resources and revenue. Following Bonaparte's interest in forests and timber, the Bourbon monarchy established the French National Forestry School at Nancy, in eastern France, in 1824.[143] Three years later the Forest Code was promulgated. Although based in large part on the 1669 Forest Ordinance, it contained significant differences. The 1827 Code levied lower fines for most infractions than had the 1669 Ordinance, and it allowed private forests to be managed by their owners rather than by the state, with the exception of a prohibition of clear-cutting.[144] Communal forests, however, did not share in the liberal protections accorded privately owned forests. They were all subordinated to the 1827 Code, which subjected them to especially rigorous rules and surveillance. The Code, in fact, cast the rural people of the communes as bad managers of the forests and "thus defined the communes as legal minors with respect to their forests, suggesting that public interest was something other than communal interest, and partially undid the gains made during the Revolution toward the recognition of communal property."[145]

The 1827 Code had in common with the 1669 Ordinance the complete prohibition of sheep and goats from the communal forests, but it allowed pigs to be pastured. Whereas the 1669 Ordinance had allowed exceptions to these prohibitions, the 1827 Code allowed almost no

Figure 3.3. The garden of the National Forestry School at Nancy. From Charles Guyot, *L'Enseignement forestier en France: L'École de Nancy* (Nancy: Crépin-Leblond, 1898). Reproduced with the kind permission of the École Nationale du Génie Rural des Eaux et des Forêts (ENGREF), Nancy, France.

exceptions, and they had to be granted by the minister of agriculture himself.[146] The new Forest Administration was more centralized than the earlier Forest Services.[147] Policing of the forest was much more strict and thorough, and many peasants as well as regional and local governments contested the new Forest Code for being too severe and too monolithic for the many different forested environments found in France. Protests soon sprang up all over the country as the people resisted the criminalization of their traditional rights of usage. Perhaps the most famous of these revolts against the new Forest Code was the "War of the Demoiselles" in the Ariège from 1829 to 1832, during which men dressed as women tried to repossess the forests and recover their use rights.[148] One of the tools of protest the rural people used was fire, especially after 1848. Fire was used so frequently and often spectacularly in ensuing years that a new law was passed in 1870 that enacted stricter prohibitions on the use of fire in the forests of the Mediterranean departments of France. This law was amended in 1883 and a new law governing fire in all French forests was passed in 1924.[149] Despite numerous additional laws such as these, the 1827 Forest Code remained the official law in France until 1953, when it underwent significant revisions.

Although forestry policy in France has deep roots in economic concerns over forest resources, timber shortages, and the maintenance of national power, evolving environmental and scientific knowledge also had important effects on its development, especially over the course of the nineteenth century. One of the most influential of these ideas was the theory that deforestation caused, or at least exacerbated, flooding originating in mountainous areas. The publication in 1841 of a study by the French engineer Alexandre Surell on the causal relationship between deforestation and flooding in the southern Alps sparked alarm and action that resulted, nearly twenty years later, in an important new law on reforestation.[150] The idea that deforestation caused torrents (the flooding of mountain streams) was not new, though. Surell built on the work of Jean Antoine Fabre, who published his study of torrents in 1797.[151] Fabre, also an engineer, believed that the primary cause of torrents and subsequent erosion in the southern Alps was the destruction of trees in the mountains. He detailed the negative effects of such flooding beyond the mountains themselves, such as the silting of harbors and damage to agricultural fields.[152] Fabre's work was widely read in France and it later influenced the founding of both the Forest Administration and the National Forestry School at Nancy.[153]

Both Fabre and Surell believed that forests could retain soil that would be lost on treeless slopes during rainfall, and that forest soils could absorb significant quantities of water. Both men thus advocated reforestation, the prohibition of cutting trees, and the elimination of grazing in mountain forests. They thought such actions in the mountains would curtail flooding, erosion, and damage to agricultural fields and harbors downstream. In the mid-nineteenth century, not long after Surell published his study on torrents, France experienced several periods of unusually heavy rainfall and severe flooding that would continue with abnormal frequency to the turn of the century.[154] These floods, pressure from the newly established forestry profession, and the need for agricultural expansion convinced Napoléon III to promulgate the 1860 Law on the Reforestation of Mountains (Loi sur le Reboisement des Montagnes).[155] This may be seen as one of the first of the forest laws to be motivated largely by environmental thinking rather than purely economic goals. It had an important impact on French forestry and led to the reforestation of tens of thousands of hectares of land in French mountains.[156] Significantly, it allowed for the expropriation of land for reforestation in the name of public utility. It was revised in 1882 to become the Law on the

Restoration and Conservation of Mountain Lands (Loi sur la Restauration et la Conservation des Terrains en Montagne).

A second environmental idea, the theory of desiccation, also had important effects on French forestry in the nineteenth and early twentieth centuries, although it had more impact on forestry in French Algeria than it did in metropolitan France. Desiccation theory posited that deforestation caused the progressive drying out of the atmosphere and the land, and, conversely, that planting trees would attract and increase rainfall.[157] This was directly related to eighteenth-century scientific ideas that related trees to atmospheric moisture and rainfall.[158] By the mid-eighteenth century these ideas were common in Europe, particularly in France and Britain. In the last half of the eighteenth century, for example, Georges-Louis Leclerc, comte de Buffon, the director of the Jardin du Roi in Paris, concluded, to paraphrase Clarence Glacken, that "a forest in the midst of scorching desert might bring rain, fertility, and a temperate climate."[159] At about the same time, the Frenchman Pierre Poivre was acting on his belief in desiccation theory and putting in place forest conservation policies on Mauritius that would have a widespread influence on environmental management in several European colonies.[160] By the mid-nineteenth century the theory of desiccation had become widely accepted throughout Europe and North America. As Marsh wrote in 1864, "I believe that a majority of the foresters and physicists who have studied the question are of the opinion that in many, if not in all cases, the destruction of the woods has been followed by a diminution in the annual quantity of rain."[161] The French scientist Jules Clavé explained in 1862 that "the terrible droughts which desolate the Cape Verde Islands must also be attributed to the destruction of the forests. In the island of St. Helena, where the wooded surface has considerably extended within a few years, it has been observed that the rain has increased in the same proportion. It is now in quantity double what it was during the residence of Napoléon. In Egypt, recent plantations have caused rains, which hitherto were almost unknown."[162] In Algeria, early scientists like Périer who thought that deforestation had caused the desiccation of the country reasoned that planting many trees would necessarily improve the climate and produce more rain.[163] Foresters believed that, even in regard to southern Algeria, "we are therefore right to conclude that the principal cause of the current dryness is the disappearance of clumps of woods which previously covered the country" and that "the rivers there have gradually diminished with general deforestation."[164] The Algerian colonist François Trottier wrote in his booklet *Afforestation in the Desert and Colonization* (*Boisement dans le désert et colonisation*) that the desert would retreat before the tree due to the "incontestable fact that the forest creates humidity and transforms the moisture regime of a country."[165] So strong were beliefs in the theory of desiccation that policies were enacted in many European colonies in Africa, Asia, and island territories to plant trees and to conserve remaining woods in order to attract rain, prevent drought, and ameliorate the climate.[166] In Algeria, as the next chapter details, it was the fear of desiccation (rather than concerns about flooding and erosion) that drove reforestation efforts for most of the nineteenth and early twentieth century. This is in contrast to metropolitan France, where the 1860 reforestation law was justified with scientific studies of flooding and erosion, as outlined above. For a quarter of a decade, from 1860 to 1885, activists in Algeria raised alarms about deforestation and desiccation partly in order to have a reforestation law passed in Algeria. As chapter 4 will show, they were ultimately successful.

Of the environmental ideas important in the nineteenth century, then, it was desiccation theory, and not the theory that deforestation caused flooding and erosion, that informed much of the development of Algerian forestry, especially with respect to reforestation policies.[167] Both of these environmental ideas, however, each of which remained influential well into the twentieth century, were seriously flawed. The theory that deforestation causes torrents was founded more on the self-interested assertions of foresters and other actors than on experimental evidence. In the words of a late-twentieth-century French reforestation expert, reforestation policies in nineteenth-century France were "born of a misunderstanding."[168] What was chiefly misunderstood was the degree to which tree leaves and leaf litter could intercept and absorb rainwater. Depending on the type of trees composing a forest and how densely they grow, research by French scientists has shown that leaves and needles can intercept up to 40 percent of water from a typical rainstorm and leaf litter can absorb up to "ten times its weight in water."[169] Although in periods of light or average rainfall trees and leaf litter may thus slow down runoff, during periods of heavy rainfall, the most common kind in North Africa, they have little effect. Moreover, the thin forests common in many parts of North Africa do not accumulate as much leaf litter as do forests in more humid zones, nor do they intercept as much water. Meanwhile, providing proof contrary to nineteenth-century theories, research conducted early in the twentieth century showed that flooding and erosion did not increase with forest removal. Several experimental studies showed that "the removal of forests do [*sic*] not of themselves increase soil erosion; rather it is the use the land is put to, following deforestation, that may lead to an increase in soil erosion."[170] These early findings on erosion and runoff have been confirmed by contemporary research.[171] This research also shows not only that tall, stiff vegetation such as trees does not impede or decrease erosion from heavy rainfall but that short, flexible vegetation, especially grasses, is most effective in reducing erosion.[172] This underscores the importance of differentiating between deforestation (removal of trees) and the complete removal of all vegetation.

As for desiccation theory, it relied on the dual premise that forests attracted rainfall and that the removal of trees decreased rainfall, lowered stream and lake levels, and led to the general drying out of the earth. The argument that the removal of forest vegetation would lead to a lowering of lake levels and stream flow and thus to a desiccated environment was never experimentally proven, only inferred from partial historical evidence.[173] It was shown to be incorrect as early as 1910, when a large-scale experiment showed an increase in stream flow when a section of forest in the watershed was removed. The moisture formerly used by the removed trees in transpiration was not lost to the atmosphere, nor did flooding increase during this and other experiments.[174] In 1911 experts concluded that changes in stream flow "are determined essentially by variations in climatic conditions which vary more in irregular cycles independent of forest conditions."[175] These experiments showing that desiccation did not follow deforestation have also been corroborated. A useful overview of the literature concludes that there is no convincing evidence that removal of forest cover reduces water yield, nor that afforestation increases water yield.[176] In fact, afforestation with some fast-growing species such as eucalyptus may significantly reduce water yield due to their great need for water.[177]

The notion that forests attract rainfall was likewise based largely on inferences and deductions from partial evidence, and primarily advanced by those with economic or ideological interests in planting trees. This idea has been dismissed in the scientific community since at least the early twentieth century.[178] Most observations of forests supposedly attracting rainfall were made in mountain forests. Rainfall in mountainous terrain, or orographic precipitation, is formed by a humid air mass ascending as it is pushed against the mountain by the prevailing winds, cooling, and condensing the vaporized water in the form of rain. Thus, mountain forests often exist because of rainfall, and not vice versa. This was one of

Pierre Poivre's misunderstandings while he was on Mauritius.[179] Yet the concept of orographic precipitation was well understood in France by the late eighteenth century, if not before. The influential French botanist René Louiche Desfontaines, who visited Algeria in the 1780s, described orographic precipitation in the Atlas mountain chain, as well as the rain shadow (dry region) that formed on the leeward side of the mountains, in this case the Saharan region.[180] This indicates that claims of deforestation and desiccation in certain geographical locations could have been countered during the nineteenth century. Moreover, long-term precipitation records show that many nineteenth-century claims of desiccation having been caused by deforestation were made during periods of unusually severe meteorological drought.[181] Such rainfall records also show that claims of reforestation causing rain were often made during periods of anomalously heavy rainfall. The claim reported by Clavé that tree plantations in Egypt had caused rainfall where none had fallen before, for instance, was made during a period of above-normal rainfall in North Africa, including Egypt, from the 1820s into the 1840s.[182]

Still, many of the French reasoned that if tree plantations had increased rainfall in Egypt, reforestation could also improve the climate of Algeria. Desiccation theory had been an implicit (and sometimes explicit) part of the developing colonial environmental narrative since the early days of the occupation, and reforestation to ameliorate the climate and attract tourists would commence in 1852 with the efforts of the army's *planteurs militaires*.[183] It wasn't until the 1860s, however, that desiccation theory significantly influenced the development of forestry in Algeria.

The original job of the Forest Service, when it was created in 1838, was to supply French troops with the wood needed for the army to function.[184] Increasingly, though, the primary motivation for the work of the Algerian Forest Service became, as in France, economic, that is, to ensure a supply of wood and other necessary forest products and to produce state revenues. To that end, the use of fire and grazing by the Algerians, both deemed destructive to forest growth, were policed and repressed. During the first several decades of French rule the 1827 Forest Code was applied to Algeria, although enforcement was geographically uneven and sporadic.[185] Fire was such a problem, however, that an additional law was passed as early as 1838 forbidding the burning of any vegetation, especially trees. Yet enforcement of forest rules remained difficult due to lack of personnel and funding. In 1843, for example, the Forest Service employed only thirty-four men for the three provinces of Alger, Oran, and

Constantine.[186] France, by contrast, had more than eight thousand forest guards at this time.[187] During the 1840s, one of the primary jobs of the Forest Service was to develop the cork oak forests for their lucrative bark; from 1844 to 1849 the service was centralized at Algiers. The government at that time was trying to exploit the cork oaks with Forest Service labor in an effort to enhance state revenues. In 1846, however, exploitation of cork oak was turned over to private enterprise in the form of concessions because the state found it too expensive to manage itself. In 1849 the service was decentralized and forest inspectors and other personnel were placed under the authority of the regional provincial governments.[188] The Forest Service was not centralized again for a quarter of a century.

Several legal developments during the 1850s had a significant impact on forestry in Algeria. In 1851, as part of a new law governing property, the government declared all forests in Algeria to be property of the state and reserved only small areas for the use of the indigenous Algerians.[189] This not only enlarged state forests but also increased state power, as it disenfranchised many Algerians from the forests they had traditionally used. Following particularly severe fires in 1853, for which Algerians were blamed, two ordinances were passed, one in 1854 and the other in 1856, to prevent forest fires. In addition, as penalty for those deemed responsible, the government actively considered instituting collective punishment of an entire "tribe" for any given fire for which a single member could be blamed. This draconian proposal would allow for the levying of fines on all members and seizure of livestock and/or property in severe cases.[190] Although not written into law until 1874, the idea of collective punishment for forest fires was already popular, and had its origins in military operations early in the occupation.[191] Meanwhile, in 1859, a decree was passed that allowed expropriation in the interests of public utility.[192] This provided the precedent for later laws in which provisions for expropriation for reforestation were particularly important, including the 1885 and 1903 forest laws discussed below.

By 1862, Forest Service personnel had grown to 266 men and enforcement was increasingly rigorous.[193] It in fact became so zealous that Napoléon III castigated the Forest Service for being too hard on the Algerians during his visit to the country in 1865. Although the emperor tried to limit the use of collective punishment and to lessen its impact on the Algerians, multiple large fires drove public opinion in favor of harsher treatment. In 1867 the government decreed that possessors of forest concessions that had suffered from forest fire since 1863 could subsequently own outright those burned sections of the concession, without any payment.[194] In addition, they were given one-third of the unburned part of the concession and were allowed to buy the remaining two-thirds at low prices. This resulted in the expropriation of Algerians from large parts of those forests under the regime of concessions, since they were assigned only one-tenth of this forested area for their traditional use. This law was reinforced with a similar decree in 1870.[195] Forest fires continued, however, becoming as much a form of protest and actual rebellion as a tool of livelihood, especially during and after the Algerian Insurrection of 1871.

About this time, many in the colony began to call for a specifically Algerian forest code to deal with such issues as forest fires that were a more serious and widespread problem in Algeria than in France. For example, 1873 witnessed truly catastrophic fires that were widely attributed to the malevolence of the Algerians. Feelings ran so high that, despite an official commission that found that the fires were not deliberately set but caused by an unfortunate series of accidents, "two of the Algerians implicated in the fires were sentenced to death and a third condemned to forced labor in perpetuity."[196] The situation culminated in the passage of a law specifically to deal with forest fires in 1874, just four years after the new law governing fire in southern France had been promulgated. This law would later become an essential part of the Algerian Forest Code promulgated in 1903. The 1874 fire law formalized collective responsibility and collective sequestration as punishment for setting forest fires.[197] It also mandated that Algerians work to help fight fires or face hefty fines and/or imprisonment. The impact of this new law on many Algerians was particularly hard: collective fines were levied and collective lands confiscated throughout the last part of the nineteenth century. In fact, the period after 1870 witnessed a substantial rise in the power of the Forest Service in Algeria, concomitant with the rise of settler (colonist) power. The Forest Service was recentralized in 1875 under the command of the Civil and Financial Affairs Department. In 1881 it was placed under the command of the Ministry of Agriculture and Commerce in France, a system called *rattachement* that gave it more power and pleased both the foresters and many colonists in Algeria.[198]

In 1885, a law "relating to the management and to the repurchase of use rights in the forests of Algeria, to the exploitation and the abuse of use rights in the woods of individuals, to the policing of forests and to reforestation" was passed in Algeria as a result of much lobbying on the part of colonists and foresters. In many ways an addendum to the 1874 law on fires, this law aimed to further curtail traditional uses of the forest. It forbade clearing of brush, for example, on the grounds that brush was simply forest that had been degraded by the Algerians' burning and grazing.[199] It was also particularly severe, especially in its treatment of those pasturing animals in the forest. Importantly, this law allowed for the expropriation of land for reforestation in the name of public utility, specifically to protect the water supply.[200] This law, too, would later be revised to form an important part of the 1903 Algerian Forest Code.

Meanwhile, the Forest Service continued to grow at an accelerated pace. Between 1862 and 1886, the number of Forest Service personnel had nearly doubled, to 493.[201] By 1898, forestry personnel had doubled again to 1,098.[202] Five years after this, the Algerian Forest Code (Loi Forestière Algérienne) was passed.

The 1903 Algerian Forest Code was based largely on the 1827 French Forest Code, but it contained several innovations that derived from the French experience in Algeria over the preceding seven decades. Drawn up by a commission created in 1892, its 190 articles included the content of the 1874 law governing fire and retained policies of collective punishment and sequestration (article 130), as well as the obligation of Algerians to help fight fires or face punishment (article 127).[203] The law of 1885 was also included in a revised form. The 1903 law was lauded by many for being simpler and less severe than the 1827 law. However, although the fines for first infractions were indeed lower, the fines for repeat offenders were actually higher than before.[204] More importantly, the authority and power of the Forest Service was extended further by this law, a fact that would have grave implications for Algerians over the next half century and continuing into the postcolonial period. The 1903 Algerian Forest Code also contained important changes reflecting the motivations underlying forestry policies and particularly in regard to reforestation. Whereas earlier efforts at reforestation were undertaken in order to try to ameliorate the climate, improve the water supply, and therefore improve public health, the 1903 law made it clear that reforestation was also to be carried out in order to maintain mountain slopes and prevent erosion.[205] Inherent in this law, then, was a belief in both desiccation theory and the idea that deforestation causes torrents and erosion. Importantly, the 1903 law allowed for the expropriation of land in state and communal areas for reforestation in the name of public utility.

By the turn of the century, then, Algerian forestry had made great strides, and Algeria was the first French colony to have its own forest code. The extent of forests was at this time estimated to be 2,816,649 hectares, of which 2,216,649 were subject to the Algerian Forest Code.[206] Armed with the new code, the more than one thousand men working for the Forest Service were a formidable force. The increase in power of the service over the last quarter of the nineteenth century can be seen in the number of infractions logged. By 1905, the number of fines levied was triple that of 1881 (22,532 versus 7,883), and revenue from them had increased at a staggering rate, nearly doubling from 1901 to 1905 alone.[207] It is worth noting that pasturing in the forest formed by far the largest single category of fines over this twenty-year period. This may have been in part due to the fact that much of the land in Algeria classified as forest land by the Forest Service was actually brush, scrub, or other land completely without trees. To illustrate, A. D. Combe, forest conservator of the province of Alger, reported that of the two million hectares of "forests" in the Mediterranean region of Algeria in 1889, approximately four hundred thousand hectares, or about 20 percent, were without trees.[208] He described this land as deforested, and undoubtedly some if it was. Much of it was also likely scrub and pasture land that could not naturally support tree growth. Two decades before this, only about 70 percent of "forest" land was estimated to actually contain trees.[209] As Charles-Robert Ageron has explained, defining unforested land as forest was a means of appropriating land that was fertile and well situated for future colonization.[210]

In any case, although the Algerians took nearly all of the blame for this presumed deforestation, a significant amount of deforestation was actually caused by the French military, the colonists, and private companies. From 1870 to 1875, for example, nearly one million cork oaks were destroyed for tannin production.[211] During the period of the Second Empire capitalist production became firmly implanted in Algeria, first in forest production and later in agricultural production.

Colonialism, Capitalism, and the Transformation of the Algerian Environment

Settler colonization by individual colonists was slowed considerably during the Second Empire due to Napoléon III's reforms in favor of the Algerians. The practice of giving free land to immigrants was halted in 1864 and the Sénatus-consulte of 1863 temporarily slowed the transfer of collective lands to colonists. Capitalist colonization, however, the sort effected by large companies and corporations, was encouraged under Napoléon III's economic liberalism and it grew significantly during this period. Among the earliest attractions for these companies were the cork oak forests, estimated at about 445,000 hectares in 1843; the potential profit from cork products, especially corks for bottles, proved irresistible for many.[212] As early as 1841 the Maisons Maillaud-Cayon et Formigli, established cork production companies with operations in Tuscany and Sardinia, requested the right to harvest and manage *all* of the cork oak forests in Algeria.[213] They were politely asked to reduce the scope of their request, but the government ultimately rejected their revised offer. Several other requests for cork oak concessions were made during these early years, but none resulted in successful operations until the late 1840s.

The first concession to result in the profitable production of cork was one of about 8,500 hectares in the Edough forests given to a businessman from Paris, Charles-Louis Lecoq, in May 1849.[214] Requests for concessions grew, as did cork profits, and by 1861 some 101,680 hectares had been awarded in concessions and a further 144,678 hectares had been requested.[215] The concessionaires had to manage the forests according to rules outlined by the Forest Service; they also had to allow the Algerians some of their traditional uses of these cork oak forests, but these were not well defined. Although the largest and most profitable concessions were awarded for cork oak forests, some were awarded in the 1860s for other species, such as olives and other oak varieties.[216] By the early 1860s, 202,000 hectares of cork forest had been awarded to thirty-four concessionaires, and these men formed a powerful lobby. Such was their power, in fact, that by 1870 they had badgered the government into practically giving them their concessions for free. As a result of forest fires in the concession forests in the 1850s and 1860s, for which the concessionaires demanded compensation from the government in the form of huge collective fines from the groups living in the forests burned, the government ultimately decided to transfer ownership to the concessionaires. This it did with the decrees of 1867 and 1870 that gave away burned sections and cheaply sold parts of the remainder of affected concessions to the concessionaires.[217] In the process Algerians lost 90 percent of their forest use rights and the government was left with only 277,000 hectares of cork oak forest, while 163,000 hectares of these forests became the private property of large companies.

Nearly 150,000 hectares of unforested land was also conceded during this period to large corporations known as the *grandes compagnies*. In 1853, the Société Genevoise, for example, was given 20,000 hectares near Setif to build villages for future colonists. To their profit, they reneged on many of their promises and did not fulfill their obligations to the state, which did not press any charges.[218] Algerians who had been expropriated from their lands were hired under exploitative conditions to farm the land for the company. Heartened by this example, other companies followed suit. The Société de l'Habra et de la Macta received 24,000 hectares to grow cotton, and the Société Générale Algérienne was awarded 100,000 hectares for a large multifaceted enterprise.[219] The Société Générale Algérienne in effect provided many contracted services for the government's colonization efforts, including work on drainage, irrigation, and reforestation as well as settling colonists and building their houses.[220]

Although immigration slowed during the Second Empire, it did not stop, and a further 233,000 French settled in Algeria and 106 new settlements were created.[221] In addition to the roughly 200,000 hectares of land turned over to companies, a further 366,000 hectares were distributed to settlers during this period, the bulk of it from 1851 to 1860, before the "Arab Kingdom" of Napoléon III was established.[222] This brought the total amount of land controlled by the settlers to 880,000 hectares, a number that would increase three and a half times by independence.[223] Nearly 1.25 million acres, then, of some of the best agricultural and forest land in Algeria was expropriated from the Algerians and awarded to French and European settlers and companies by 1870. It was during this period that the capitalist production that defined agricultural development in colonial Algeria began. Despite the image that many liked to create of rugged smallholders taming the land, the majority of agricultural development was in fact carried out by corporate landholders and large private landowners.

Much of the land that was turned over to colonists and companies had been expropriated from the Algerians by various means since the beginning of the occupation. Land was lost by Algerians as a result of fleeing military assault, collective punishment, and the laws of 1844 and 1846 on vacant or waste land. The land law of 1851, which codified the government's ownership of the forests, also provided the legal backbone for the policy of cantonment, which was used to acquire much collective land during the Second Empire.[224] Cantonment assumed that the Algerians had more land than they needed or could use "productively," and offered to recognize and give legal title to part of the "tribal collective lands" if they would cede the part of their lands they did not need. In practice, though, the policy served as a means of confiscating land. At least 61,000 hectares of the best land was taken from sixteen groups between 1857 and 1861.[225] Many of these groups lost between 40 and 85 percent of the most fertile of their lands, which led to a diminution of their livestock since much of their land had been used for pasture. Colonel Lapasset of the Arab Bureau lamented that around Orléansville cantonment had completely ruined "'the most beautiful tribe . . . diminished it by half' after having 'taken for the colonists all the fertile land.'"[226] Cantonment, he wrote, condemned the Algerians "to the most shocking of miseries."[227] To add insult to injury, many of the colonists and *grandes compagnies* rented the Algerians' own land back to them at exorbitant rates instead of farming it themselves. Cantonment terrorized the Algerians, and although the Algerian government, with the enthusiastic backing of the settlers, wanted to extend it indefinitely, the French metropolitan authorities ended what they saw as veritable plunder in 1861.[228]

The final way land was garnered for colonization during this period was via the Sénatus-consulte of 1863. This piece of legislation was signed by Napoléon III as a way to reassure the Algerians that not all of their land would be taken from them (which is how most Algerians saw cantonment functioning). With the aim of stopping expropriations, it declared the "tribes" the owners of the lands they had traditionally occupied. The end result, however, was the transfer of about 14 percent of all rural collective land in Algeria to colonization by 1870.[229] These were of course the best, most fertile lands (and forests) that were lost to the Algerians. In the productive northern Tell zone, the application of the Sénatus-consulte succeeded in extracting approximately 80 percent of collective lands, which were then transferred to colonization efforts.[230] This was accomplished by demarcating the boundaries of the traditional land declared property of the "tribes," and then dividing it into *douars*.[231] Douars were administrative regions (basically villages) that were governed by a council that "owned" the land for the members of the douar. The land was divided up among the members of the douar and could be further subdivided into individual, privately owned parcels. This essentially turned collective land into a commodity to be easily bought and sold. Furthermore, the douars were drawn up to include people and land from different lineage groups. This was a deliberate action on the part of the French to "reduce the influence of the chiefs and to break up the tribes."[232] Arguing that the nefarious actions of the Arab nomads needed to be curtailed, one member of the legislative committee exclaimed, "born in the deserts of Arabia and Asia, everywhere they move, this race has made the desert around it. It is this race that . . . obliterated the last vestiges of Roman civilization."[233] Those involved in formulating this legislation were thus certainly aware of the colonial environmental narrative and used it to justify the new law. A military officer summarized the impact of the Sénatus-consulte by writing that it was the "the most efficient war mechanism that one could invent against the native social structures. It was the strongest and most resourceful instrument that was put in the hands of our settlers."[234] Nonetheless, most settlers believed that the Sénatus-consulte was too generous to the Algerians. Their vigorous lobbying resulted in its suspension in 1870.

By that time, the Algerians had lost 80 percent of their land in the Tell region, the most agriculturally productive land in Algeria, and nearly 15 percent of their lands in Algeria overall. Only 3.5 percent (73,946 hectares) of forest land, or land that had been defined as forest, was left for Algerian use.[235] For a traditional economy that depended heavily on pastoralism, this loss of land and resources was devastating. As outlined earlier, pastoralism, especially in arid and semiarid regions, requires large amounts of land that are grazed only infrequently to take advantage of the local ecology. This also requires that livestock and people be able to move frequently, usually over long distances. The precolonial management systems of collective lands were accommodated to this kind of extensive livestock production. The cantonment of the Algerians, however, and the further loss of collective land with the 1863 Sénatus-consulte, circumscribed not only land but also movement, since the creation of douars included boundaries. The loss of their traditional pasture lands forced many pastoralists to settle, a trend that would be accelerated in the coming decades.[236] For those Algerians who depended on forest-based agropastoralism, the loss of nearly all their forest land was equally devastating.[237]

It was not only the use of the Algerian environment that changed during this period. The environment itself changed as grain production and other commodity farming spread and the forests were turned to cork and timber production through the cultivation of "proper" trees. Meanwhile, most of the traditional modes of production, and therefore survival, had been criminalized and severely curtailed. These included not only pasturing livestock in the forest but also the collection of firewood, food products, medicinal plants, and other forest resources. By 1870 many, if not most, rural Algerians were angry and terrified for their very lives.

Their situation was made worse by a period of drought, harvest failures, famine, locust invasions, and disease that resulted in the deaths of three hundred thousand Algerians in the late 1860s.[238] Faced with growing settler power in the last few years of the Second Empire, and fearing even worse conditions in the future, the Algerians revolted in the chaotic aftermath of the Franco-Prussian War and the change of government it precipitated in Algeria. The new Third Republic installed a colonialist government in Algeria whose policies would have profound impacts on the colony over the next several decades.

As colonization spread in Algeria and the settlers gained resources and power during the Second Empire, the declensionist colonial environmental narrative matured into its complete form. What began as a tale of the untapped fertility of the Algerian soil, wasted by the improvidence of the "natives," was transformed into an accusation of the ruin of the land by the hordes of Arab nomads and their herds that for centuries had destroyed the environment. Although evident in its entirety in only a few of the more vitriolic commentators of the time, parts of narrative, and much of its spirit, were found in key developments of the period. The policy of cantonment, for example, relied on the premise that land that did not appear to European eyes as if it were being used was actually vacant and ownerless. This was either a profound misunderstanding of the indigenous mode of extensive pastoralism and the use of fallow, or it was a willful misinterpretation. By the end of the Second Empire, the use of the declensionist narrative is evident in the writings Henri Verne, François Trottier, and others. This mature narrative would practically become a mantra during the Third Republic. During that period, the narrative was not only rhetoric from angry settlers and others; it also became a tool for changing laws and policies in ways that further disenfranchised the Algerians.

Plate 1. *The Remains of a Roman Aqueduct in the Region of Cherchell (Ruines d'aqueduc romain près de Cherchell),* 1868, by Victor-Pierre Huguet (1835–1902). This painting, set in Algeria, is characteristic of many romanticized nineteenth-century representations of Roman ruins in North Africa. Such paintings often included the figures of "nomads" such as those represented here. The ruined aqueduct, the scant vegetation, and the nomads and their camels all work together to hint at the French colonial narrative of a ruined landscape, as in similar orientalist work of the period. Cherchell was previously the large Roman town of Caesarea. Its aqueduct was one of the ruins of Roman waterworks most frequently mentioned during the colonial period. From Musées des Beaux-Arts de Rouen, France. Photograph by Catherine Lancien/Carole Loisel, © Musées de la Ville de Rouen. Reproduced by permission.

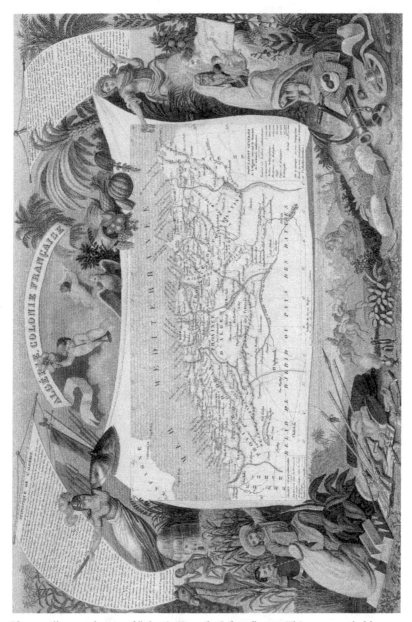

Plate 2. Illustrated map of "Algeria, French Colony," 1847. This map, probably produced for an illustrated atlas of France, depicts the "southern region," that is, Algeria, early in the colonial period. The sketches surrounding the map represent Algeria as a veritable cornucopia of agricultural products. Grains, fruits, vegetables, rice, salt, and dates figure prominently in the iconography. The Algerians appear in subordinate positions, apparently being enlightened by the French. It clearly promoted a colonial position and invoked the environmental fertility narrative common during the early years of occupation. The open manacles and broken chains assert that the French have liberated the Algerians from the shackles of slavery and oppression. From the collection of the author.

Plate 3. *Kabyle Herders (Bergers kabyles),* no date, by Eugène Fromentin (1835–1902). This painting, by one of the most influential of France's orientalists, places Berber herders and their plump sheep in a lush, idyllic, pastoral setting, reflecting the commonly accepted belief that the Berbers were good environmental stewards compared to the destructive Arabs. From Musée du Louvre, Paris, France. Photo by Réunion des Musees Nationaux/Art Resource, New York. Reproduced by permission.

Plate 4. *Threshing Wheat in Algeria (Décipage des blés en Algérie)*, 1853, by Adolph-Pierre Leleux (1812–91). Leleux, a lesser-known orientalist, painted this work before the colonial environmental narrative was widespread in Algeria. It recalls threshing wheat in rural France, where horses were often used to loosen the grain from the chaff. Such images helped to normalize life in Algeria at a time when settlers were still being recruited to colonize the territory. From Musée des Beaux-Arts, Lille, France. Photo by Erich Lessing/Art Resource, New York. Reproduced by permission.

Plate 5. *Pilgrims Going to Mecca (Pélerins allant à la Mecque)*, 1861, by Léon Belly (1827–77). Created just as the declensionist environmental narrative was beginning to gain widespread use, this is one of the earliest desert landscape scenes to garner popular acclaim. Although the focus of the canvas is on the nobly depicted people and camels, hints of the colonial narrative are found in the bare and desiccated landscape. From Musée d'Orsay, Paris, France. Photo by Réunion des Musees Nationaux/Art Resource, New York. Reproduced by permission.

Plate 6. *Le Sahara,* 1867, by Gustave Guillaumet (1840–87). Considered one of the key paintings of the orientalist genre, Guillaumet's *Le Sahara* illustrates the colonial environmental story of decline just as this narrative was becoming ubiquitous in Algeria. A landscape almost entirely denuded of vegetation surrounds a dead camel, the center of the tableau. The collapsed rider and camel apparently left to perish as a small band of survivors departs in the distance suggest that the environmental destructiveness and brutality of the nomads' way of life must end. From Musée d'Orsay, Paris, France. Photo by Réunion des Musees Nationaux/Art Resource, New York. Reproduced by permission.

Plate 7. *The Land of Thirst (Au Pays de la soif)*, 1869, by Eugène Fromentin (1820–76). Taking its title from the last sentence of Fromentin's book narrating his travels in Algeria, *The Land of Thirst* reiterates a very common sobriquet for Algeria from the 1860s well into the twentieth century. This painting, depicting Algerians dead from thirst, reinforces the colonial story of desertification. In painting the desert as a ruined and desiccated landscape, Fromentin, like Guillaumet, helped to naturalize the colonial environmental narrative, primarily for the French public. From Musées Royaux des Beaux-Arts de Belgique, Brussels, Belgium. Photograph © Royal Museum of Fine Arts of Belgium, Brussels. Reproduced by permission.

Plate 8. The entrance to the Algerian Palace at the 1889 Universal Exposition in Paris. In the leftmost archway lies the sarcophagus erected by the Ligue du Reboisement to the memory of Algeria's forests. From Glucq, *L'Album de l'exposition* (Paris: Ch. Gaulon, 1889), planche 75. Reproduced with the kind permission of the Center for Southwest Research, University of New Mexico, Albuquerque, New Mexico.

CHAPTER 4

The Triumph of the Narrative, 1870–1918

THE FALL OF NAPOLÉON III'S SECOND EMPIRE thrilled the settlers and the civilian administration in Algeria. The Third Republic that was born in France in 1870 was very supportive of the settlers and their goals. The new French government helped to crush the Algerian Insurrection of 1871–72 and allowed the subsequent repression of the Algerians. In the words of an eminent French historian, "the French of Algeria henceforth imposed their will almost without opposition."[1] Under the Third Republic settler power grew, capitalist agricultural production spread, and the Algerians experienced further losses of property and resources. The declensionist environmental narrative that had matured during the Second Empire triumphed with the settlers, many of whom wielded it as a tool for implementing change. This colonial narrative was used, for example, by Auguste Warnier to justify the 1873 property law named after him as well as the 1887 law that subsequently modified it. Others invoked the narrative when arguing for instituting private property and sedentarizing nomads. It was also employed extensively by many associated with forestry in the colony, especially those involved in lobbying for changes in the forestry laws and for reforestation. It was used widely to promote the planting of eucalyptus, a tree introduced from Australia that some thought would bring wealth and rain to Algeria. Paulin Trolard, a physician and influential colonist, used the narrative along with many settlers and foresters to lobby for significant changes in the laws related to forestry in the colony. It thus facilitated, and was partially incorporated into, the 1903 Algerian Forest Code. During this period, the narrative also began to influence the science of botany in Algeria, although it would not be widely embraced by botanists and ecologists until after World War I.

The Third Republic, Insurrection, and Algeria Assimilated to France

The Second Empire fell on 4 September 1870, to the delight of the French in Algeria, when Napoléon III capitulated two months after having declared war on Prussia and two days after being captured at the Battle of Sedan. Within days a provisional government was established in Paris and a republic declared. Paris remained under siege, however, until January 1871. Elections were held despite this, and a government was assembled within months. Although the war with Prussia officially ended that spring, an uprising, the Paris Commune, turned Paris once again into a battleground. The "insurrection turned revolution" was brutally crushed by the new government headed by Adolphe Thiers. The number of dead among the revolutionaries (*communards*) is estimated to have been between ten and thirty thousand.[2] Over the next several years different parties jockeyed for power, and though many wanted to reestablish a monarchy, the republic was maintained. By 1875, the Third Republic was firmly established with the passage of its constitutional laws; it would remain in place until World War II. During this period, little progressive social legislation was passed in France and the gap between the rich and the poor continued to grow, fueling discontent among the lower classes. The Third Republic did introduce secular education, and in 1882 free and compulsory primary schooling was established under Prime Minister Jules Ferry.[3] The relatively prosperous early 1870s gave way to economic depression in the late 1870s, and economic depression plagued France and much of Europe for the next two decades. During the 1880s protective customs tariffs were reintroduced in an effort to support French producers. For the most part, however, the government maintained the status quo and privileged property owners over other members of French society.[4]

Under the terms of the Treaty of Frankfurt (10 May 1871), the recently declared German Empire annexed the provinces of Alsace and Lorraine, and France paid a war indemnity of five billion francs.[5] For the next several decades France would be preoccupied with Germany. Many believed that in addition to securing its borders against Germany, France needed to raise its birth rate to compete demographically. Some saw expansion overseas as a way to increase the size, power, and population of France and thereby keep Germany and other European rivals in check. Many also believed that colonial expansion would benefit France economically. Not only would colonies provide cheap or free raw materials for industrial production, but they could also provide markets for manufactured goods. According to Prime Minister Jules Ferry, "the establishment of a colony is the creation of a market."[6] In the late nineteenth century, as European capitalist production boomed, markets were perceived to be more and more essential.[7] By the 1880s, overseas expansion was routinely justified in France by asserting that it was good politically, economically, and culturally. Many French also believed that their civilization was superior and that they therefore had a duty to spread their civilization and culture to lesser peoples in foreign lands. This was the so-called civilizing mission (*mission civilisatrice*).

Although France had been slowly expanding its overseas territories since 1815, when nearly all colonial possessions were lost as a result of the Napoleonic Wars, acquisition accelerated during the 1880s.[8] Tunisia was conquered easily in 1881, the result of an exaggerated border incident and of agreements between European powers at the Berlin Conference of 1878. Expansion into sub-Saharan Africa, specifically French West Africa and French East Africa, began in earnest in the 1880s, and domination of French Indochina was firmly established in the 1880s as well. Overseas expansion continued well beyond the nineteenth century, however, and Morocco was made a French protectorate in 1912 after years of jostling with other European powers, primarily Germany. Much of the impetus for this expansion came from the French colonial lobby, which grew and developed significant power during the 1880s. An informally organized group of colonists, businessmen, geographers, and explorers, the colonial lobby consisted in its early form as many different committees, each focused on a particular colony or group of colonies. By the 1890s these disparate groups became more formally organized. For instance, the Committee for French Africa was founded in 1890. In 1892 a "colonial group" was established in both the Chamber of Deputies and the Senate. In 1893 the French Colonial Union was formed, composed of more than four hundred French companies with colonial interests. Several chambers of commerce in cities such as Marseilles also lobbied on behalf of colonial interests.

One of the most influential members of the colonial lobby was Eugène Étienne. Born in Algeria, the son of a military officer, Étienne won a seat in the French Parliament as a deputy from the province of Oran in 1881.[9] A member of Parliament for nearly forty years, he founded the colonial group in the Chamber of Deputies and served as undersecretary of state for the colonies, minister of the interior, and minister of war. He was representative of the growing number of French and other Europeans who had been born on Algerian soil and considered Algeria to be their country. By the mid-1890s, the French and Europeans who had been born in Algeria outnumbered those who had immigrated. The European population in Algeria had doubled between 1872 and 1896, from roughly 280,000 to 578,000.[10] In 1872, it owned 765,000 hectares of agricultural land; by 1875, another 446,000 hectares of collective land had been confiscated as punishment for the 1871 Algerian Insurrection, and half of this had been given over to colonization efforts.[11] Those Algerians identified as participants in the 1871–72 insurrection were also punished with monetary fines in the amount of about sixty-five million francs, which amounted to roughly 70 percent of the capital of those fined.[12] Within just a few years after the fall of Napoléon III and his "Arab Kingdom," the Algerians had been crushed, the Bureaux Arabes disbanded, and the settlers were firmly in control. As a result of the toll of warfare, famines, epidemics, dislocations, and the disruption of the traditional economy, the Algerian population reached its nadir around 1871. An estimated six hundred thousand Algerians died between 1861 and 1871 alone. The Algerian population had fallen from an estimated three million to about 2.1 million in the forty years since Algeria had been occupied by France.[13] An unknown number of deaths from the insurrection (estimated in the thousands) occurred between 1871 and 1872. The Algerians were, in numerous ways, too weak to resist the onslaught of the settlers and the changes they demanded over the next several decades.

The governors-general in the 1870s and 1880s were nearly all in favor of colonization, including Admiral de Gueydon (1871–73), Albert Grévy (1879–81), and Louis Tirman (1881–91). Governor-General Chanzy (1873–79), formerly an officer in the Bureaux Arabes, was recalled at the insistence of the settlers for being too "pro-Muslim" and not facilitating the settlers' goals.[14] These two decades saw the extension of the civil territories in which many French laws were applied. The Algerians were doubly taxed, with the system of *impôts Arabes* (Arab taxes) as well as direct taxes, but they saw little benefit from it since the French mayors of the communes arranged the budgets to benefit the settlers. Algerian representation in government bodies was decreased, a harsh new penal code solely for indigenous Algerians was implemented in 1881, and their movement was circumscribed by requiring them to obtain a permit to leave their douar.[15] By 1881, the Tell region was composed of communes in which the Europeans had complete control. The average commune at this time had 294 Europeans and twenty thousand Algerians. The taxes paid by this large indigenous population supplied the vast majority of the communes' budgets, giving rise to the common saying that "the communes live by eating the *indigène*."[16] This repressive taxation continued through World War I, and by 1909 the Algerians were paying about half of all taxes in the colony although they owned well below half of the land and capital.[17] The Arab taxes were not abolished until 1919. In 1881, an important administrative change took place that "attached" most of the ministries and services in Algeria (such as the Forest Service) to their counterparts in France, effectively bypassing the governor-general and the Algerian government. Since the government in Paris was usually too busy or indifferent to what was going on in Algeria, this system of rattachement, which lasted until the late 1890s, effectively increased settler power even further. During the tenure of Governor-General Tirman (1881–91), for example, members of the French government who strongly supported the settlers in Algeria, such as Eugène Étienne, "became the real masters of Algeria."[18]

After this triumph of the settlers, official colonization was resumed, and the government once again began distributing free land to colonists. One of the first such measures was the allocation of 100,000 hectares of land to refugee families from Alsace and Lorraine displaced by the peace agreement between France and Germany. Although that effort was ultimately a failure, nearly 600,000 hectares of land were distributed to European settlers between 1871 and 1890, bringing the total amount of agricultural land controlled by the settlers to 1,635,000 hectares by 1890.[19] Much of this land was expropriated from the Algerians with a new legal mechanism that was implemented in 1873, the "Warnier Law." The Warnier Law of 26 July 1873 in essence retooled the 1863 Sénatus-consulte, so reviled by the settlers, in order to explicitly dismember indigenous collective property and transform it into individual private property.[20] This law abrogated parts of the 1863 law so that the latter no longer protected collective properties but instead subjected most transactions of land in Algeria to strictly French laws. This allowed undivided collective property to be alienated from joint owners if only a single owner could be persuaded (or bribed) to sell their "portion." Furthermore, the 1873 law dictated that unused or "unproductive" lands be placed under the authority of the office of colonization, which resulted in the confiscation of much grazing and fallow land. It is estimated that in this way approximately 377,877 hectares of land were transferred from Algerian to European owners by 1890.[21] A complementary law was passed in 1887 that required land not already surveyed and "delimited" to be surveyed and delimited. Called the "little Sénatus-consulte," this law, applied well into the twentieth century, broke another 337 "tribes" into 529 douars by 1934. Under it, countless deeds of property were issued for individual shares of jointly held land.[22] This provided the means for allocation of a further 976,795 hectares to the state domain by 1899, of which 682,989 hectares were classified as forest land.[23]

Until this period, French colonization in Algeria had not been a resounding economic success. In the 1880s, however, several developments combined to create the conditions for an economic transformation of Algeria. Up to this point, wheat, which did not require large amounts of capital to produce, had been the staple of colonization, and it continued to be produced over ever-larger expanses of land into the twentieth century. The economic value of the cereal harvest soon was eclipsed, though, by Algerian wine production. Viticulture spread quickly during the 1880s thanks to the increased availability of credit in Algeria, a drop in international prices for wheat, and the phylloxera crisis that ruined many vineyards in France. As a result there was a phenomenal increase in vineyards in Algeria over this decade, from 15,000 hectares in 1878 to 110,000 in 1890.[24] Production was soon dominated by large landholders and by 1935 vines covered 400,000 hectares and wine produced 50 percent of the colony's export earnings.[25] The spread of the vine fueled the need for land at the same time that it concentrated landholdings and furthered commercial production of nonsubsistence crops.

By the early 1890s, two reports had revealed to the government in Paris the despoliation occurring in Algeria since the settlers had come to power two decades earlier. The revelations were so shocking that Governor-General Tirman was asked to resign and a special commission of inquiry into the situation was appointed, headed by former French Prime Minister Jules Ferry. The commission's report condemned the methods of the

Algerian administration, the policy of *refoulement* of the Arabs, "the sequestration of their lands, the harshness of the forestry regime, and the greed and racial discrimination displayed on the part of the French community."[26] Despite the appointment of a new governor-general, Jules Cambon (1891–97), who tried to implement reforms, little change was made. The colonialist lobby in Algeria and in Paris succeeded in blocking nearly all of his efforts. Cambon did succeed in 1896 in abolishing the system of rattachement, attaching the ministries to Paris, but was soon after relieved of his post at the insistence of Étienne and others in Parliament. In 1898 and 1900 two important decrees were promulgated. The first created a new governmental body called the Délégations Financières (Financial Delegations). The second decree gave Algeria independent civil status. The Délégations Financières was composed of about two-thirds European settlers, who were elected, and about one-third Algerians, who were partly elected and partly appointed from a very small eligible group. Responsible for the budget of the colony, which was usually rubber-stamped by the government in Paris, this group of men held significant power, since the creation of the Délégations Financières effectively gave budgetary autonomy to Algeria.[27] It retained substantial power over the next several decades, until the rumblings for independence began in the early 1950s.

In 1901 Charles Jonnart was appointed governor-general of Algeria, a position he held until 1911 (with a short absence from 1901 to 1903). Jonnart, too, aimed to reform the administration of the Algerian population but in the end failed to overcome the protests of the settlers and the colonial lobby. He did succeed, however, in stimulating the Algerian economy, which boomed in the prewar years.[28] Charles Lutaud succeeded Jonnart as governor-general (1911–18). Lutaud was thoroughly supportive of the settlers and their goals and did not pursue any reforms that might be beneficial for the Algerian population. The leadership in Algeria, then, from 1871 through the First World War, was either supportive of settler power or blocked by the settler and colonial lobbies from making any substantial reforms. Land acquisition by European settlers continued, as did official colonization aided by the government. The government itself had also gained a significant amount of land for the public domain using the "little Sénatus-consulte" of 1887.

By the end of the First World War, the settlers controlled 2,123,288 hectares of prime farmland (about 28 percent of all arable land), nearly three times what they had controlled in 1872, and 194,159 hectares of forest.[29] The state held an estimated five to six million hectares of land, of which about 2.5 million were classified as forest land.[30] In all it is esti-mated that at least 7.5 million hectares of the best agricultural and forest land had been expropriated from the Algerian population by World War I.[31] At this time, only 1,967,955 hectares owned by Algerians were being cultivated (by a population at least six to seven times larger than the set-tler population) due to their loss of land over the preceding decades. Be-cause of increased population growth (4,741,000 by 1911), per capita agri-cultural production was declining sharply. Squeezed into smaller and smaller areas of land and taxed mercilessly, more and more Algerians were forced into wage labor. Indeed, by 1914 wage labor "had become one of the most characteristic features of colonial agricultural production."[32] Thirty-two percent of rural Algerians were working as sharecroppers by 1914.[33]

The property laws passed during the Third Republic were thus respon-sible for a large gain in land by both settlers and administration that im-miserated vast numbers of Algerians during this period. The law that laid the foundation for these property laws, the 1873 Warnier Law, provides a particularly compelling example of the use of the declensionist environ-mental narrative to justify and facilitate a singularly important facet of the colonial project.

Settler Power, Private Property, and "Fixing the Nomads to the Soil"

Auguste Hubert Warnier (1810–75) was the principal author of and spokes-person for the property law passed on 26 July 1873. Warnier was at that time a member of the French Parliament, having been elected as a deputy from Algeria in 1871.[34] A veteran colonist, Warnier had arrived in Algeria in 1834 as a member of the Scientific Commission. Also a physician, Warnier later served as a military surgeon and as an Arab Affairs officer for a few years. By 1848 he had been appointed director of civil affairs in Oran and in 1870 he was named the prefect of Algiers. A militant for settlers' rights, he wrote numerous books and articles, many of them coauthored with his friend and colleague Jules Duval, a former judge.[35] Warnier has been described as the "most articulate and effective colonialist spokesman of the era."[36] He was a polemicist who was decried by some but who had many enthusiastic supporters among the colonists and in the colonial lobby. He fought vigorously for years in his various capacities to increase the settlers' wealth and power, and the passing of the 1873 property law was in many ways his crowning achievement. This law, as noted above, provided a mechanism to dismember traditional collective property, transform it into saleable individual private property, and make it available for colonization. It also mandated that lands not being used "productively," according to European agricultural norms, were essentially vacant and thus belonged to the state. Historians have pointed out that this law was conceived as a reaction against the 1863 Sénatus-consulte, which the settlers perceived as wasting perfectly good land on the "natives." It has been less well recognized that this law, and the larger agenda held by Warnier and settler partisans of which it was a part, invoked the declensionist environmental narrative for much of its justification.[37]

In addition to agitating for the conversion of collective property into private property, Warnier and most settlers were pushing for other changes. They wanted the administrative organization of the country, the tax system, and the judicial system overhauled to better support the "assimilation and civilization" of indigenous Algerians. Just as they wanted to see all land transactions governed by French law, they wanted French law to govern most other aspects of life in the colony. Some partisans, like Warnier, went to great lengths to spell out how and why these changes must be made.[38] Warnier's publications provide a clear example of justificatory arguments that were widely used among the settlers and in the colonial lobby. In his publications Warnier also clearly made generous use of the declensionist environmental narrative.

In *L'Algérie devant l'empereur* he based his argument on the proposition that the Algerians were made up of two fundamentally different groups, the Arabs and the Berbers. Warnier was a strong adherent of the Kabyle myth. The Berbers were the original peaceful, industrious, sedentary inhabitants of North Africa. The Arabs, on the other hand, were descended from "hordes of nomads from Arabia" who had conquered North Africa in the eleventh century, "a devastating torrent, destroying, pillaging, slaughtering everything in their path."[39] The Arabs had never ceased to be "elements of disorder" since their society was "nothing but the legacy of the Barbarian invasion."[40] Warnier concluded from these "historical facts" that the Arab nomads were a threat to the security of the country and should be sedentarized.

Warnier blamed the Arabs for environmental destruction down through the centuries, from the time of the Hillalian "invasion." The land where the "Arab tribes" lived was "long ago a sort of terrestrial paradise that legends say was covered in groves [of trees] and rich pastures, irrigated by abundant waterways. Today this land is a sterile desert, bare and without water, that all call the *land of thirst.*"[41] With his alarmist style, he even warned that "France itself, so prosperous, would soon become a desert if it were in the hands of the Arabs."[42] He explained further that the Arabs didn't have any "bonds with the land": they lived in tents, moved from place to place, and abused the land. Warnier attributed their burning and overgrazing to their system of land tenure, accusing them of living on the land "communally with all the members of their tribe . . . without ever giving it [the land] the least care."[43] The Berbers, though, Warnier praised because he believed they were strongly attached to the land. They already had private property, and "the titles of property even mention the number of trees of each species."[44] Thus, "in Algeria one can, without mistake, be sure that where the land is desolate, without trees, one is in Arab territory; on the contrary, where there is beautiful cultivation, beautiful trees, woods and forests, one is in Berber territory."[45] The implication was that forcing private property on the Arabs could make them better environmental stewards and increase their agricultural productivity. However, since "the Arabs have a horror of working [and] the land becomes sterile in their hands," it would be better "to oblige the Arabs to contain themselves to limited areas and to sell the rest [of their land] to newcomers who know how to restore life where death reigns."[46] The solution, for Warnier, was the expansion of private property throughout Algeria, and that required a new law. Most of Warnier's publications contain very similar statements, reasoning, and conclusions.[47]

Warnier was a key member of the original group assigned to study the issue of property in the colony and draw up a draft law in 1871. He was closely involved in every stage of the draft's development and participated energetically in the debates over it in the Parliament. In 1872 a parliamentary commission was ordered to study the draft law and Warnier headed its work. Warnier shepherded the law through several revisions and debates, in which some high-ranking critics, including Governor-General Chanzy, raised their voices against the harshness of the law.[48] Despite the critics, the law was passed quickly, without discussion or amendment, on 26 July 1873; as Charles-Robert Ageron has said, "Warnier had the last word."[49]

Warnier had brought his arguments—in favor of the institution of private property, and against the purported wastefulness of leaving land in the control of the nomadic Arabs—to his work on the law and the debates over it. He proclaimed in his parliamentary report that "it is in the double goal of returning Algeria to its ancient productive power by a better assignment of property and putting a stop to the unequal repartition of land between the natives and those who have immigrated from France or Europe that has led the government to request . . . a law constituting private or individual property where it does not exist."[50] It is clear that Warnier's ultimate goal was to secure more property and associated resources for the colonists. In this he succeeded, although he did not live long enough to enjoy the results of his labors. He died in 1875, but the Warnier Law would remain in effect until 1892, when its application was suspended in response to the Ferry report.

The 1873 Warnier Law, and the environmental narrative embedded in it, formed the foundation upon which subsequent property laws, especially that of 1887, were built. This can be seen in the work of the commission that was appointed in 1881 to consider its revision in the face of growing criticism of the Warnier Law and the "ruin of the Arab element" it had engendered.[51] The result, with the help of veteran settler parliamentarians like Eugène Étienne, was instead a new law that "reinforced and extended the rights conferred on Europeans by the Warnier law."[52] The 1887 law mandated the resumption of the surveying and delimiting of property and the issuing of land titles. It thus was a continuation of the 1863 Sénatus-consulte, without any protections for indigenous communal lands. For this reason it was frequently called the "little Sénatus-consulte." This law set the stage for breaking up collective land on a large scale and substantial land acquisitions by the government. By 1899, the lands surveyed and delimited since 1887 left only about 10 percent of communal lands to their original owners; the rest, 3,370,800 hectares, was taken by the state or defined as private land.[53] Since most of the Tell region had been surveyed and delimited by 1900, in the twentieth century the 1887 law was applied mostly in the southern territories, and it was particularly damaging to the collective lands of the nomads. By 1921, some 3,114,792 hectares of the best land had been appropriated by the state.[54] Of this, 1,179,664 hectares were classified as forest land.

Much of the land acquired for colonization south of the Tell was put into livestock and cereal production. Colonial wheat fields, for example, increased from 158,607 hectares in 1872 to just under 1,200,000 hectares in 1910.[55] A large portion of this increase occurred in the south thanks to the introduction of the North American technique of dry-farming. Based on a three-year cycle of plowing, planting, and fallow, dry-farming frequently made cereal production possible in areas of low rainfall.[56] This allowed cereal production to spread over large parts of the High Plateaus as a speculative venture by large landholders. Over time, however, this technique desiccates soils in arid areas, and thus it essentially "strip-mined" the soil, contributing to "the impoverishment of the soil and hence to the decline of agricultural production."[57] In areas of pasture land in the arid High Plateaus, dry-farming also disrupted the local vegetation and made the plowed areas unsuitable for grazing, sometimes for years, until the native vegetation was reestablished.

The loss of pasture land hit the nomadic pastoralists of the High Plateaus particularly hard. The number of sheep owned by Algerians decreased by half from 1870 to 1911, down to only about 6.25 million.[58] These devastating losses were not due to disease or drought. In fact, for much of this period there were above-normal rains. During this same period, the number of sheep owned by European colonists in the region doubled.[59] European colonization centers had expanded onto the High Plateaus since 1871, and the best grazing and agricultural lands were transferred from the Algerians to the Europeans via the 1873 and 1887 property laws.[60] The migrations of the nomads had been regulated (and greatly restricted) on an annual basis by the government since 1863 as a result of article 34 of the 1863 Sénatus-consulte.[61] The loss of collective grazing land, the loss of "forest" land, and the restrictions on migrations took a profound toll on the Algerian nomads and their herds. Whereas about 60 to 65 percent of the population were estimated to be nomads at the time of conquest, by 1911 the figure was down to only 18.5 percent.[62] Growing numbers of nomads were forced to sedentarize, mostly due to poverty. Many sold their animals and their newly constituted private land to settlers or to indigenous landlords out of desperation, leading to a few wealthy families in each area becoming owners of large herds. Others migrated seasonally in order to work on European agricultural estates as wage laborers, one of the few sources of income remaining to them.[63] Still others worked for the large commercial companies harvesting alfa grass in the High Plateaus. Harvesting and export of alfa had begun in the 1860s and by the turn of the century many areas in the High Plateaus were denuded of this hardy perennial grass.[64]

Given these significant changes, in 1901 Governor-General Révoil ordered a study of the nomads in the southern regions, which were then military territories. He ordered another such study in 1902 and Governor-General Jonnart later ordered a third study on the commerce of the southern oases. In 1906 a book was published by an influential geographer, Augustin Bernard, and Napoléon Lacroix, a captain in the Indigenous Affairs Service. In *The Evolution of Nomadism in Algeria* they attempted to synthesize all of the recent information on the nomads that the military had provided in their many separate reports to the governors-general.[65] The authors also included liberal amounts of their own writing and interpretation. Their goal was to try to explain the changes that had occurred and to assess the possibilities for the future. The product, however, is more an extended justification of the changes that had taken place, and the declensionist environmental narrative is deeply embedded throughout their writing. Many of their arguments are reminiscent of Warnier's a quarter of a century earlier. Like Warnier and many others, they blamed the "Arab invasion" of the eleventh century for ruining North Africa. Citing Ibn Khaldoun they claimed that "the Arabs hurled themselves across North Africa, like a cloud of locusts, spoiling and destroying everything in their path. . . . They cut down all the trees. . . . Civilization was ruined and the country was changed into a desert."[66] They concluded that "arborescent vegetation and agriculture certainly lost ground during the Arab invasion and the following centuries. . . . The Arabs have been fatal . . . by their way of life and their habits . . . it is their sheep, their camels, their goats that have ruined North Africa. . . . The Berbers have helped them very much in this [devastation]."[67] Unlike Warnier, then, Bernard and Lacroix did not invoke the "Kabyle myth." The Berbers were also blamed for overgrazing and burning the forests down through the ages, although they were not mentioned extensively in this work.

Despite the large numbers of nomads that had been sedentarized and the great reduction in herd size by 1906, Bernard and Lacroix persisted in highlighting what they claimed was environmental damage caused by the nomads and their herds. "The natives manifest a veritable hatred for trees," they wrote, and because of overgrazing the "forest passes to the state of maquis [scrub]."[68] They declared that "nomadism, with its herds, tends to enlarge its domain unceasingly, to sterilize bigger and bigger regions. . . . Insecurity favors their progress."[69] They strongly favored sedentarization and approved of its occurrence so far in the region. In fact, they prescribed as much sedentarization as possible, explaining that wherever agriculture might be viable, "the nomad must disappear before the [settled] agriculturalist."[70] Fully subscribing to the common perception that Rome had successfully colonized a then fertile North Africa and made it a granary, they culled the Roman past for examples of how to deal with the nomads. They noted that "during the period of calm and prosperity . . . of Roman domination, there was certainly a retreat of nomadism. . . . Massina tried to . . . attach them to the soil. . . . They [the Romans] sedentarized the natives with security and the promise of a better quality of life; their dwellings were grouped into villages."[71] As a warning to the administration, perhaps, they noted that the Romans were not completely successful in sedentarizing all the nomads and that, when troubles arose, "the natives returned to their ancient state."[72] Eternal vigilance and surveillance, they implied, were needed to keep the unruly nomads under control. Sedentarization was thus made to seem a tested method successful under Roman domination.

Bernard and Lacroix also shared the belief in desiccation theory so common in Algeria at this time. They were careful to explain that the consequences of deforestation, "the deterioration of the climate, the water supply . . . in one word the total ruin of the country, are the inevitable results of the disappearance of the vegetal cover of the soil"[73] Because they identified grazing in the forest as the primary cause of deforestation, they recommended that it be forbidden and that communal pastures should be improved to accommodate the livestock of the Algerian pastoralists. In fact, the government had been making efforts to improve the pastures and watering points in the High Plateaus for a decade or more. A Pastoralism Service was established in 1896 to oversee and coordinate the activities of various departments, including Veterinary Medicine.[74] Bernard and Lacroix, though, remained fixated on the forest, of which there was very little in the southern territories. In their conclusion they wrote that the true issue in the region was the equilibrium between the forest and pastoralism. The forest, they avowed, "must always come before all else."[75]

Despite such grand pronouncements, these authors were more cautious than many others in the colony who wanted to reforest practically the entire country, including the High Plateaus, at any cost. Bernard and Lacroix cautioned in their last sentence that "the pastoral industry must not be ruined to try to reforest regions that are not suitable to it."[76] Many other colonists in Algeria, however, dreamed of reforesting the High Plateaus and creating vast new centers for European colonization. One of them even believed that the desert itself could be conquered and the nomads sedentarized by reforesting with eucalyptus trees.

The Rise and Fall of Eucalyptus

During the 1860s and 1870s a veritable mania for planting eucalyptus trees (principally *Eucalyptus globulus*) swept through Algeria. Within a decade

Figure 4.1. Eucalyptus trees lining a street in Algiers just outside the casbah. From Clément Alzonne, *L'Algérie* (Paris: Fernand Nathan, 1937), 61. Reproduced with the kind permission of Armand Colin.

of the first extensive plantings in the 1860s, hundreds of thousands had been planted (see figure 4.1). A fast-growing tree, eucalyptus was praised for many different reasons. It was proclaimed that eucalyptus would ameliorate the dry climate, improve public health, supply Algeria and France with wood, and allow colonization to succeed in the desert. Reforestation in general was widely believed to be necessary for colonialism to survive in North Africa, and the declensionist environmental narrative was of course quite useful to many of those making such claims. It was particularly evident in the writings of the man who did the most to spread the word and the seeds of eucalyptus in Algeria, François Trottier, the apostle of eucalyptus.

Eucalyptus globulus was introduced into Algeria by the French businessman Prosper Ramel, who had become infatuated with the tree during a trip to Australia. Ramel procured seeds from Ferdinand von Mueller, director of the botanical garden at Melbourne, and brought them back to France in 1854.[77] A short while later he took seeds to Algeria. Ramel was convinced that eucalyptus trees could cover the mountains, "purify the swamps, chase away the fevers," and provide cigarettes to replace "the stupefying fumigations of hashish."[78] The earliest experiments in the propagation of eucalyptus were successfully carried out in 1861 and 1862 by Auguste Hardy at the Jardin d'Essai in Algiers with seeds provided by Ramel. A year later colonists began to introduce the exotic trees. Some of the early adopters included the Count of Bellerôche, A. Cordier, Armand Arlès-Dufour, and Charles Bourlier.[79] The bulk of these early plantings were carried out in the Mitidja region, a swampy area south of Algiers. Ramel himself grew eucalyptus at Hussein-Dey.[80] Most of these proprietors hoped that the eucalyptus plantations would help to dry out the swampy areas and purify the air of the miasmas believed to cause the fevers that had killed so many in Algeria.[81] They also planted them for agricultural windbreaks and to provide wood for construction and fuel.

The largest of these early plantings to be carried out by a settler was that of François Trottier. Between 1867 and 1876 Trottier planted nearly forty hectares on his properties at Hussein-Dey, Maison Carré, and Fondouk in the Mitidja. He was, according to one of his contemporaries in France, "seized with the fever of eucalyptus."[82] During this time Trottier wrote several short books, in addition to reports, on eucalyptus and the urgent need for reforestation generally in Algeria, the earliest in 1867. He is regarded as the popularizer, "the apostle" of the exotic tree and he is largely responsible for its widespread adoption in the colony.[83] His efforts were very successful. Over the course of the next decade, "it was by the hundreds of thousands that eucalyptus were planted and the names of Ramel and Trottier were on everyone's lips."[84] Indeed, Dr. Eugène Bertherand, member of the Algiers Departmental Council of Hygiene and Public Health, estimated that by 1876 about 1.5 million eucalyptus trees had already been planted in the three provinces of Algeria.[85] In 1878, a booklet written for the Universal Exposition in Paris claimed that about four million had been planted in the colony.[86] Shortly following the 1867 publication of Trottier's book, *Notes on Eucalyptus and Subsidiarily the Necessity for Reforestation in Algeria* (which enjoyed a second edition soon thereafter), the Société Algérienne planted 100,000 eucalyptus around Lake Fetzara in the province of Constantine, and the Compagnie Franco-Algérienne planted 135,000 in the plains of the Habra and Macta.[87] In addition to individual colonists and commercial companies, the military, municipalities, and government road and bridge engineers (Service des Ponts et Chaussées) planted eucalyptus on a large scale for their needs.[88] Eucalyptus was expected to reduce intermittent fevers (malaria) among the workers. It would also provide wood for construction, telegraph poles, railway lines, mineshaft supports, naval vessels, and firewood. So successful was eucalyptus in Algeria during this period that one 1876 essay on the subject remarked that "a stranger who was not instructed of its exotic origin would take it for one of the indigenous trees of the region."[89] Twenty years later a physician who owned a farm south of Algiers noted that most villages had groves or avenues of eucalyptus.[90] Trottier's propagation and promotion of eucalyptus were deemed such important contributions to Algeria that he was awarded the coveted Cross of the Legion of Honor in 1878.[91]

Trottier was born in France in 1816 and traveled to Algeria with the military in 1839. Once out of the army, he remained as a colonist, raised livestock, and later produced linen and cotton. He became mayor of the commune of Rassauta in the 1850s, during which time he contracted malaria.[92] By the 1860s and 1870s, Trottier had become well known and fairly influential. He served as mayor of Hussein-Dey and in the early 1870s as the president of the Society for the Encouragement of Reforestation in Algeria, which he helped to establish.[93] Both his publications and his achievements were noted and quoted in the literature on Algeria into the 1930s.

Trottier declared in the book mentioned above that "eucalyptus would be the great product of Algeria" for three primary reasons.[94] First, eucalyptus would provide France with the wood it needed. Second, it would purify the country, regularize the rains, attenuate the winds and floods, ameliorate the climate, and thus favor civilization. Finally, this improvement of the climate would prevent the moral deterioration commonly attendant upon living in Africa.[95]

This 1867 text contained a long section on the deforestation of Algeria that repeatedly invoked the declensionist environmental narrative in the form developed by this time.[96] Trottier explained that in the distant past North Africa was clothed with dense forests. The subsequent forest destruction had been caused by Arabs who burned trees and allowed their herds to overgraze. He also implicated the common-property system of the Arabs as one of the reasons for deforestation, claiming that common property is incompatible with planting or conserving trees. The deforestation that had occurred had changed the climate and thus the moral state of the inhabitants. The Arabs, in his opinion, were "a race that successively diminishes to make room for another more civilized [the French] and of a higher level."[97] For Trottier and many others, reforestation was a solution for a multitude of perceived problems, from the torrid climate to the "degenerate natives." For reforestation to work, though, Trottier specified that severe regulation of Algerians' fires and grazing would be needed, and he insisted that the French reforestation laws as well as the Corsican fire edicts be implemented in Algeria.[98] In another book, *Afforestation and Colonization: The Role of Eucalyptus in Algeria,* published in 1876, Trottier repeated his call for the application of the Corsican fire edict.[99] He also demanded collective punishment (possible since the 1874 fire law was passed, but not implemented until 1877) of entire groups when they were found guilty of causing forest fires, including deporting the "guilty tribes" from the Tell region altogether. He condemned the 1863 Sénatus-consulte and the government's "sentimentalism" toward the Algerians, concluding that because Arabs destroyed forests, they were a plague (as they had been throughout history), and civilization must destroy them.[100]

Trottier was one of the first in Algeria to promote reforestation as a way to extend colonization into the more arid, desert areas of southern Algeria. In his 1867 booklet, *Afforestation in the Desert and Colonization,* Trottier wrote that the desert would retreat in front of eucalyptus plantations since forests of eucalyptus would create humidity and transform the climate.[101] He instructed the reader that "colonization in the interior (southern Algeria) is not possible except after planting trees."[102] Trottier also predicted in 1876 that reforestation would sedentarize the nomads and produce a settled population.[103] After identifying the Sahara Desert as the great enemy, he noted that rainfall had been diminishing in Algiers for twenty years and suggested planting forests to create a "dam of another sort" that would, moreover, provide "a source of revenues."[104] The idea of planting trees to serve as windbreaks and to humidify the country was not new, having been proposed and implemented to a degree since early in the occupation.[105] Trottier, though, was one of the earliest to articulate the idea of "green dams" in the desert, an idea that would later captivate colonial and postcolonial governments alike throughout Africa and the Middle East.

In 1875, at the age of fifty-nine, Trottier applied for and received from the government eighty-four hectares to plant with eucalyptus over a period of twenty years, providing that he planted and harvested at specified rates and that he respected the Forest Code. This land was provided without charge from the state's domain. The Algerian government supported his application to the president of France because it agreed that plantations of eucalyptus in the valleys of Isser and Sebou, "exceptionally fertile but unhealthy and denuded," would purify the countryside and be of incontestable utility.[106] Furthermore, the government projected, the forest plantations "will give, in the near future, because of their rapid growth, wood in sufficient quantity for the needs of other Europeans installed nearby"; they were thus "works of public utility."[107] His project was supported by glowing testimonials from the prefect of Algiers and the head of the Forest Service, who agreed that it would improve the climate and contribute to the public good. Trottier expected to make a considerable profit from his eucalyptus plantations, on the order of about 6,500 francs per hectare every nine years, a very considerable sum for the time.[108] It is unclear from the archival record if Trottier ever planted many eucalyptus trees on this land, but he remained a militant activist for reforestation and joined the Reforestation League (Ligue du Reboisement) when it was formed a few years later.

By the late 1870s, however, problems were beginning to be identified with eucalyptus, and the predictions of great profit made by Trottier and others were not being fulfilled. By the end of the century eucalyptus was denigrated by many for being "a tree without use and even harmful."[109] Specifically, the wood of eucalyptus was hard to cut and mill since it cracked and splintered easily. It also appeared that eucalyptus stunted the growth of other plants, especially trees, and could even kill them.[110] More worrying for many was the fact that eucalyptus soaked up so much water from the soil that it caused several springs to dry up.[111] Even Trottier was forced to destroy some of his eucalyptus plantations because he believed that they were drying out his wells.[112] By the 1890s the rage for eucalyptus in Algeria had waned considerably, and eucalyptus is hardly mentioned in the comprehensive *Forests of Algeria* written by Henri Lefebvre, head of the Forest Service, in 1900. It was not even much used for reforestation by the Forest Service, which preferred Aleppo pine and cork oak, both of which had proven commercial value.[113] Near the end of the colonial period, just prior to World War II, Paul Boudy wrote that in Algeria eucalyptus was considered "without true economic interest" and that it covered only "a few hundred hectares in total."[114]

Nonetheless, the history of the rise and fall of eucalyptus in Algeria provides an interesting window through which to examine a particular use of the declensionist environmental narrative by one person over a period of time. François Trottier, the "apostle of eucalyptus," invoked the narrative in order to justify his eucalyptus plantations, argue against communal property, and curtail the traditional livelihood activities of the Algerians. Importantly, he used it to successfully argue for free land from the government, since, as he explained, the government could and should provide free land to planters in order to increase the tree cover of Algeria.[115] It is clear from his publications and the archival record, though that Trottier's primary goal was economic. He believed that he would become rich by planting and selling eucalyptus. He hoped that the fast-growing eucalyptus could replace the lucrative but slow-growing oak. He cited, as an added benefit, the fact that the land planted in eucalyptus would be free for crops after twenty years since this was the period of time needed to mature trees for harvesting.[116] In doing this he revealed that his main interest was economic and not climatological, ecological, or even patriotic (supplying Algeria and France with wood).

Many of the ideas Trottier expressed in his writings, including his rendition of the declensionist environmental narrative, met with approval during the colonial period. He was cited, for example, in the catalog on the forests of Algeria prepared for the 1878 Universal Exposition in Paris and seen by large numbers of people.[117] Trottier was also cited frequently by Dr. Paulin Trolard, founder and president of the Ligue du Reboisement de l'Algérie, who referred to him as "father Trottier." Trottier, who had already (with the government's approbation) helped to establish the Society for the Encouragement of Reforestation in Algeria, became an early member of Trolard's Ligue in 1882. Many of Trottier's ideas, especially the declensionist environmental narrative, were carried forward by the Ligue du Reboisement which became an extremely powerful lobbying group from 1882 to 1915.

Forestry and the Role of the Ligue du Reboisement

The League for the Reforestation of Algeria (La Ligue du Reboisement de l'Algérie) was formed in 1881. This benevolent-sounding organization was established by Dr. Paulin Trolard, who directed it for twenty-nine years.[118] Based in Algiers, the Ligue quickly drew a broad membership. The six hundred members counted in January 1882 increased to fifteen hundred by October of the same year. The first issue of their monthly bulletin proclaimed that their single goal was to reforest Algeria in order to combat drought.[119] The second issue contained the "Statutes of the Ligue," in which the Ligue declared that, among other things, it would "keep up a continual agitation about the reforestation question," and that it would work to obtain from the administration "desirable modifications to forestry legislation." Employing a desiccationist discourse, Trolard and his Ligue fought relentlessly to prevent deforestation and advocate for reforestation. The "Call to Algerians" in the first issue set the tone for much of the Ligue's subsequent work. "Every deforested country is a country condemned to death. . . . Algeria is a deforested country. The few forests that the teeth of the herds have spared, the incendiary Arab threatens to take from us in a few hours. . . . [I]f . . . we decide to fight until our climate is transformed [by reforestation], it will be wealth, it will be life, it will be Algeria returned to its original fertility: it will be Algeria becoming the granary of France!"[120] If no action was taken, Trolard warned in a Ligue publication the next year, "the Sahara, this hearth of evil, stretches its arms toward us every day; it will soon enclose us, suffocate us, annihilate us!"[121] Most of the members of the Ligue over the next thirty years shared these views. Throughout its existence, the Ligue enjoyed the membership and support of an important and influential group of people. From its inception it counted high-ranking foresters, government officials, professors of agronomy and botany, physicians and veterinarians, colonists, and wealthy landowners among its members.[122] In 1905, even Governor-General Jonnart joined the Ligue. Because of its clear use of the declensionist environmental narrative, its influence in the area of forestry policy and law, and its wide-ranging membership, the Ligue provides a useful lens through which to examine the influence the colonial narrative came to have during this period.

Among the founding members of the Ligue were several forest subinspectors, including Messieurs Chitier, Gravier, Tisserand, Wendling, and Forest Inspector Reynard. Reynard, Tisserand, and Wendling served on the central committee of the Ligue in various capacities and were very active in the organization.[123] Chief Algerian Forest Conservator M. Mangin was also a member of the Ligue, as was one of the three departmental forest conservators, M. Combe.[124] The director of forestry in Tunisia joined in 1905, and forester Paul Boudy joined in 1910. Foresters were some of the most active and influential members of the Ligue during its existence. The Forest Service used the Ligue to further its own ambitions in several ways, and the Ligue did not hesitate to use many of these government foresters to facilitate its mission.

Earlier, especially during the period of the "Arab Kingdom," tension had existed between the Forest Service and many in the military and the administration who thought that the rigorous enforcement of forest rules created problems and unrest among the Algerians. Their primary goal was to maintain peace and security. Thus, some in the military sought to protect the Algerians from the Forest Service and its zealous agents.[125]

These tensions persisted and were in many ways exacerbated in the period after 1871, because the Forest Service continued to gain power and to enforce forest rules and extract punishments for infractions more and more rigorously. While the numbers of fines levied in civilian areas climbed steeply in the 1870s and 1880s, the numbers levied in the military areas remained modest.[126] There were outspoken critics of the Forest Service and the way the military treated the Algerians. One of these, Commandant P. Wachi, warned that "if we dispossess the natives of these rights [to the forest], we compromise the future of the colony and we will make of Algeria an Ireland, where the religious hatred that already divides the people will be exacerbated by terrible social hatred."[127] He also pointed out that if the nomads were deprived of property they would become nomads and therefore pose a greater security threat. Wachi rebutted the declensionist narrative, stating that it was inexact to blame the Algerians for deforestation and drought. He argued instead that the "diminution of the immense glaciers that advanced, several centuries ago, in the valleys of Europe" caused the atmosphere to dry out in Algeria and the forests to disappear.[128] Although there were a few other critics, the declensionist environmental narrative had become ubiquitous in Algeria by this time, and the Forest Service continued its harsh application of the forest laws to the approbation of most colonists and many administrators. A primary impetus for the increasing severity in the application of forest rules during this period was the 1872 report by the director of forestry in France, Louis-François Tassy, on the Forest Service of Algeria. Tassy bemoaned the degraded state of Algerian forests and set out guidelines that became a template for the reorganization of the service and the enforcement of existing forest rules.[129] The system of rattachement (1881 to 1896), which attached many of the government departments, including the Forest Service, to their counterparts in Paris, increased the Forest Service's power further as it bypassed the governor-general of Algeria and the civil administration. The Forest Service also benefited from the work of the Ligue as the latter raised public awareness of the problem of deforestation and the need for reforestation.[130]

Raising public and governmental awareness of the imminent dangers of deforestation, desiccation, drought, desertification, and ruin, and thus highlighting the need for reforestation, was one of the Ligue's foremost goals and activities. It was commonly proclaimed in the pages of the *Bulletin de la Ligue* that colonists would soon be forced "to flee in front of the invasion of the Sahara" and that they had to choose "between life and death."[131] From 1881 on, the Ligue had newspaper articles written, published numerous booklets and brochures, and promoted public conferences by its members on the dangers of deforestation and the urgent need for reforestation. Dr. Trolard proudly let it be known that, "thanks to this agitation, I have received numerous letters [of support] from all over."[132] Soon after its inception, the Ligue began to capitalize on this heightened awareness as it petitioned local and national governmental bodies for financial support as well as for legislative changes promoting reforestation efforts.[133] Three of its earliest demands put to Governor-General Tirman were for prizes for reforestation projects, to be given at the regular agricultural competitions among settlers; a new forest law for Algeria that would include the 1882 French law on the conservation of mountain lands; and an increase in forestry personnel in the colony.[134] Tirman immediately expressed his eagerness to implement the prizes, and in 1883 prizes for reforestation were decreed by government order.[135] In fact, Governor-General Tirman acted as an activist for reforestation, and forest protection in general, during most of his administration. He believed that deforestation had been occurring "for centuries" and that Algeria "threatened to become infertile due to drought, a fatal consequence of the disappearance of our forests."[136] With his help, the other two demands of the Ligue were also met within a few years.

The Ligue remained a powerful pressure group over the next three decades, and tailored its tactics according to the nature of its various goals.[137] In order to increase the number of forest agents, for example, it petitioned not only the governor-general but also the General Council of Algiers, a governing body that advised the governor-general. This approach was successful, and in 1882 the General Council voted to demand a "strong augmentation of forestry personnel" and an increase in the budget for forestry projects.[138] Forestry personnel and the budget were increased throughout the 1880s and the 1890s. The Ligue lobbied the elected Superior Council of Algeria, the top governing body below the governor-general, several members of which were also Ligue members. A majority in the Superior Council was disposed favorably to the Ligue's rhetoric and one member warned that, "if we aren't careful, in a few years, our lands completely denuded, . . . the Algerian Tell will become an extension of the Sahara."[139] They promoted reforestation as well as more rigorous application of the forest laws.

The Ligue also lobbied in France and frequently petitioned the Parliament in Paris. Trolard contacted the minister of agriculture in Paris,

M. de Mahy, many times on behalf of the Ligue. In 1883, the Ministry of Agriculture in Paris instated a new policy that provided forest plants (tree seedlings) to colonists at no cost.[140] In June 1886 it also approved Trolard's application for the Ligue to be recognized as an "establishment of public utility," which entitled it to certain subsidies by the government.[141] Senators Étienne and Le Lièvre were instrumental in obtaining this recognition.[142] Being thus recognized also allowed the Ligue to accept the gift of twenty thousand francs left to it by member Dr. Eugène Bodichon at the time of his death in 1885. This was a significant amount of money and funded much of the Ligue's early work. Closer to home, the Ligue also took on the role of a sort of intermediary policing body. It intervened with the prefect, for instance, to support the complaints of colonists against other colonists and against Algerians for various "abuses" of the forest.[143]

The Ligue lobbied for four years to have the 1882 French law on the conservation of mountain lands (essentially a reforestation law) applied to Algeria. Beginning in early 1882, the Ligue repeatedly sent requests to the governor-general that this law be applied in the colony.[144] This was part of the Ligue's general agitation for a new forest law to be created specifically for Algeria. Up to this point the 1827 French Forest Code had been applied with varying results, supplemented by the 1874 Algerian fire law. The original study resulting in the fire law had contained a second half that was not passed into law at that time. Many like Trolard and other Ligue members felt that action needed to be taken to promulgate these other recommendations as law. In July 1882 the central committee of the Ligue sent a letter to the French minister of agriculture that contained draft legislation (*projet de loi*) based on the second half of the 1874 study, while also recommending incorporation of the French 1882 reforestation law so as to make it applicable to Algeria.[145] A few months later, the Ligue sent more letters pleading for the institution of a new forest law in Algeria to both the governor-general and the Superior Council. The Superior Council voted in October 1882 to make a formal statement on reforestation. The council announced that it had determined that, since the water supply in the country was threatened by deforestation, it must "instantly call the governor-general's attention to the necessity and urgency of taking measures to promptly remedy this alarming situation." It requested that the governor-general budget ten thousand francs per year for each of 1883 and 1884 to address the crisis. Finally, it made sure "to thank the Ligue du Reboisement" for the documents it had provided to the council and for its "devoted help on a topic that strongly interests all of the colony."[146] It is clear from this, as Trolard noted in the *Bulletin de la Ligue,* that the Ligue was having a marked influence at high levels of government. It would continue to do so. In 1886, for instance, an Algerian forestry group was formed in the French senate (Groupe Forestier Algérien au Sénat), due largely to the increased attention that the Ligue had brought to bear on Algeria's forest situation.[147]

Due at least in part to the Ligue's agitation, Governor-General Tirman was persuaded of the urgency of the problem of deforestation, the need for reforestation, and the benefits of the 1882 French reforestation law. He ordered a study of the feasibility of reforestation in Algeria by the Forest Service in February 1884, "with a view to the eventual promulgation in Algeria of the law of 1882."[148] This study, the *Programme générale du reboisement,* resulted in the development of the December 1885 forest law. Dr. Trolard was ecstatic. He wrote in the *Bulletin de la Ligue,* "the forest law is finally passed!! The project that we supported has been adopted."[149] This 1885 law included much of the original 1874 draft law that was never passed, including restrictions on the use of forests by Algerians. It also allowed the "buying back" of Algerians' use rights (especially grazing rights) where the forest was deemed in need of protection and curtailed the rights of European property owners to cut trees and clear land where it damaged the forest.[150]

New in the 1885 law, however, was a section on reforestation. Article 13 allowed expropriation of land for reforestation upon a declaration of public utility. For example, reforestation of a site might be considered necessary to protect the water supply. The land that could be expropriated included not only mountain land (thus including the elevated plateaus to the south) but also dunes, and indeed any land deemed necessary to protect water sources or the public health. Thus, this section was designed "to act as a substitute for the promulgation of the [1882 French law on the conservation of mountain lands], the integral application of which would have been impossible in Algeria and would not respond to its needs. . . . [I]t was especially the protection of springs/streams that article 13 envisioned."[151] Finally, the 1885 law provided even more detailed restrictions on the use of fire. The documents pertaining to this study and to the promulgation of the 1885 forest law, in which reforestation was an essential element, reveal much about the prominence and use of the declensionist narrative during this period.

Governor-General Tirman, in his letter of instruction ordering the forest inspectors to complete the reforestation study that resulted in the

1885 *Programme générale du reboisement,* made clear that the goals of re-forestation in France (to control torrents and erosion) were not the same as those in Algeria. By this time, desiccationist discourse had become widely accepted in Algeria, at all levels of government, by scientists, foresters, and academics, and among the general public. Tirman explained in his letter that, in Algeria, it was a question of "regularizing the discharge of springs and waterways, of creating a barrier against the southern winds, and, finally, of tempering the ardors of a scorching climate. Reforestation in Algeria must be conceived as giving the country the water it lacks in summer."[152] He echoed Trottier as he advocated planting a "thick curtain of forests" to "protect the Tell" from the encroaching Sahara. The foresters who contributed to the final report all agreed with Tirman's assessment, and made clear in their reports that deforestation was widespread and that it was the main cause of the dry climate in Algeria.[153]

These foresters had varying ideas, however, concerning the best course of action for their particular provinces. The forest conservator of Oran, Auguste Mathieu, for instance, wrote that conservation of existing forests and scrub would be more effective than a large program of reforestation.[154] Moreover, he concluded that the French law of 1882 would not be applicable to most of his territory and was therefore of little interest. The conservator of Constantine, M. Calinet, concluded that reforestation was necessary only in some parts of his territory, but that in these locations it was an urgent matter. The 1882 French law would therefore be applicable to those particular parts of his territory. He warned, though, that the state must harvest more of the cork from its forests in order to pay the high costs associated with reforestation.[155]

Regarding the province of Alger, Forest Inspector J. Bert was quite restrained in his opinions. Although he acknowledged that reforestation would be useful in many areas, he believed that private initiatives, such as those encouraged by the Ligue du Reboisement, would likely suffice.[156] His superior, Conservator Combe, however, argued that in the areas where reforestation was critical, such as in the south of Alger Province, state action would be needed. As to methods, he concluded that while reforestation was necessary in some areas, conservation of existing trees and scrub would suffice to increase forest cover in many others. To do this, he argued, required an even more rigorous and strict enforcement of the forest laws along with state acquisition of more forest land. The state, however, had been acquiring land easily for decades, since, in Combe's words, "most often the natives only possess collective usage rights to land and it is easy to classify it as vacant and thus in the state domain by virtue of the law of 16 June 1851."[157] Nonetheless, Combe explained in detail that the existing 1882 French law on the conservation of mountain lands was mostly inapplicable in Algeria because, among other reasons, it did not allow for the expropriation of land not yet degraded; that is, it did not go far enough. He recommended that the new law for Algeria should "accord to the state the right of expropriation of mountain lands the restoration or the afforestation of which is necessary for the alimentation of springs and the regularization of their discharge."[158] Thus, as a whole, the report argued for a more vigorous and strict application of existing conservation measures and forest laws and for expanding the state domain, both of which were mostly contained in the 1874 projet de loi that was never passed, and for a modified application of the 1882 French law on reforestation. As A. D. Combe acknowledged in 1889, the forest law passed in December 1885 "was the practical result of [this] study on reforestation."[159]

Although all of the reports described the destructive habits of the contemporary Algerians (the "ravages of the natives"), only one of the four foresters, Calinet, discussed the historical causes of deforestation or mentioned the ancient past of North Africa. He mentioned several places that "were forested long ago" and stated that North Africa, "the granary of Rome, possessed vast forests [with] large wild animals" that had disappeared with deforestation.[160] He alleged that burning and grazing were responsible for this destruction down through the ages, but did not directly blame the Arab invasion for past deforestation. The lack of discussion of the past in this document is likely due to the rather constraining letter of instruction from Governor-General Tirman. He clearly ordered a study of current conditions in the colony as well as the potential benefits of a general program of reforestation, including the possibility of implementing the 1882 French law on the conservation of mountain lands. He specified in regard to the Tell region that the report should also contain "general considerations on the future of agriculture and industry, the value of the land, the structure of the country, the nature of the soil and subsoil, the altitude of the plains and the mountains, the extremes of temperature, [and] the discharge of springs and waterways at different times of year."[161] Regarding the High Plateaus and the deep south, he specified additional work. This degree of detail in the instructions constrained most of the authors from making additional comments, including on the historical causes of deforestation. In his report, Conservator Mathieu actually

complained about the amount of work and the short time allowed to prepare the study.[162]

The most important and high-ranking of the foresters contributing
to the report, A. D. Combe, conservator of Alger, published another work
the same year that did elaborate on the historical causes of deforestation
and made full use of the declensionist environmental narrative. Combe
began his *Notice sur les forêts* with a description of the thick, shady forests
that covered North Africa "from Tangier to Tripoli" during antiquity, and
then blamed the "Arab invasion and the pastoral practices of the new oc-
cupants" for the subsequent deforestation.[163] He mourned the loss of the
ancient fertility of Algeria, the former "granary of Rome," and castigated
the Arabs for their "fires followed by grazing that are still today the worst
element of destruction for the forests of Algeria."[164] Near the end of his exposition, Combe disclosed one of his primary goals: increasing the amount
of "forest land" under the control of the state and therefore the Forest
Service. He explained that "in these deforested countries, the small remnants of these ancient forests must be conserved as precious and to this
end placed immediately in the hands of the state either by exchange or by
expropriation."[165]

In these pages Combe, a member of the Ligue du Reboisement since
1882, extolled this organization and its influence. He wrote that "public
opinion has been very occupied for several years with the dangers that the
disappearance of the forests could have for Algeria, where the influence of
woods on the climate, on streamflow, and on public health is more considerable than in any European country. A League for Reforestation has
formed in the three departments [provinces], having as directors intelligent and devoted men who endeavor by their enlightenment, their advice,
and their encouragement to propagate the ideas of conservation and protection that we owe our forests."[166] Combe shared with Trolard and other
Ligue members a firm belief in the colonial environmental narrative of
decline and put it to specific use in his writings and in his work. He used
it explicitly to try to justify an expansion of the land under the control of
the Forest Service. He appears in fact to have served as a key link between
the Ligue and the administration due to his elevated status in the Forest
Service. It is worth noting that Combe also blamed some recent deforestation on European settlers and the military, especially in the south.[167]

Most foresters during this period, and especially those associated with
the Ligue du Reboisement, invoked the declensionist environmental narrative to justify various positions, plans, and policies in the general realm

of forest management. In 1880, for example, Forest Inspector Reynard wrote of the area around Laghouat, in southern Alger Province, that "it was at an earlier time highly populated: many Roman ruins cover the country. . . . The rivers there have gradually diminished with general deforestation. This idea is corroborated by the numerous traces of ancient forests."[168] Since the Roman period, Reynard explained, the Algerians, especially the nomads, had been burning, overgrazing, and cultivating in the forests; little by little, their actions had "created the sand dunes, where all vegetation has disappeared."[169] Reynard used the declensionist narrative to create a sense of urgency about what he considered the degraded state of the southern section of Alger Province, roughly from Boghar to Laghouat. He dreamed of improving the pastures and restoring what he believed to be the remnants of ancient forests. He wanted to settle the nomads, "fix them definitively to the soil," and build a railroad across the Sahara.[170] He complained that the state had abandoned "this vast country . . . to the whims of nomads without suspecting that they would ruin it completely."[171] Reforestation, would bring the "regularization of water discharge [and] would impart to much of this land an exceptional fertility and, at the least, completely restore these vast pastures."[172] Most of his project was based on the exploitation of recently discovered underground water supplies. Wells, he believed, would provide water around which the nomads could be settled. This in turn would "make available for European colonization vast expanses [of land] where pastoral production could be immediately developed, so important for the future of Algeria."[173]

Like Combe, Reynard used the narrative primarily to justify the acquisition of land and resources. Both of these men generated crisis narratives that raised public awareness and facilitated new legislation, in this case the 1885 forest law. Indeed, five of the ten supporting documents sent by the Ligue with its projet de loi to the Minister of Agriculture in 1882 were reports (with very similar alarmist content) written by Reynard. As a result of new legislation (including the 1873 Warnier Law), increased public awareness, and the increased power of the settlers through the civil administration of Algeria, the forest domain grew from about two million hectares in 1872 to nearly 3.25 million hectares in 1888.[174] The vast majority of forest land in 1888 was under the control of the civil administration and the military; some was privately owned; and the least amount was held as communal forests for the Algerians.[175] This precipitous rise in the amount of forest land under the control of the state was accompanied by a steep rise in fines and punishments for forest infractions. One senator in Paris wrote that "the citations shower on [the Algerians] like hail."[176] The number of citations for forest crimes more than doubled between 1881 and 1888; the revenue also increased from 1,265,312 francs in 1884 to 1,658,958 francs in 1890, most of it wrested from the Algerians either in cash or as corvée labor ("work days").[177] The increase in revenue further enhanced the power of the Forest Service during this period. Indeed, during the 1870s, and especially the 1880s, the Forest Service became in the words of one historian a "veritable State within a State."[178] Despite the increased severity and rigor with which the forest laws were enforced, the growth of land classified as forest land, and the augmentation of the power of the Forest Service, the Ligue continued to agitate for still more action to save Algeria's forests. To draw attention to its cause, it placed a funerary monument "erected to the memory of the forests of Algeria" at the 1889 Universal Exposition in Paris (see plate 8). The Ligue hoped that visitors would "appreciate the severe but just judgment that presided over the building of this dismal sarcophagus situated to the left of the entrance of the Algerian exposition."[179]

During the 1890s, primarily under the administration of Governor-General Cambon, things changed. The government in Paris began to take notice of what many were calling abuses of Algerians under the civil regime, especially by the Forest Service. An investigation was launched and the result was the 1892 "Ferry Report," which condemned much of the civil administration and singled out the harshness of the forestry regime.[180] As a result of this attention, the long, drawn-out process of developing an Algerian forest law resumed. The report also led to the end of the system of rattachement in 1896, which restored to the governor-general power over the various departments, including the Forest Service. Governor-General Cambon was able to moderate to a certain degree the harshness with which the forest laws were implemented, although he did not succeed in significantly reforming the administration in Algeria as he had hoped to do. The Forest Service saw this as a check on its powers and was quite critical of the report.[181] The Ligue du Reboisement also criticized the report and lobbied Ferry and other senators in Paris before and after the report was published.[182] By the end of the 1890s, constantly challenged by Cambon's efforts to decrease the stringency of forest law enforcement, the Forest Service, like many of the colonists, was agitating for a new comprehensive forest law for Algeria. One had been presented in the Parliament in Paris as early as 1893, but it was delayed and modified repeatedly over the next several years, and wasn't approved until 1903.[183]

In 1900 the chief forest inspector in Algeria, Henri Lefebvre, wrote a comprehensive volume, the first major official text on the colony's forests in a decade. In his chapter on the "actual state of the forests and the causes of destruction," Lefebvre featured a particularly complete version of the declensionist environmental narrative, including references to Ibn Khaldoun.[184] He lamented the loss of the thick forests of antiquity and primarily blamed the nomadic invaders, who with "their considerable herds led to the destruction of the forests."[185] A few pages later Lefebvre made it clear that in Algeria, due to the alarmingly degraded state of the forests, either a new forest law was needed or the 1827 French Forest Code had to be implemented fully. The colonial narrative, for Lefebvre, was a useful way to focus attention on deforestation and the necessity of passing the new forest law then under debate and revision in the highest levels of government in Algiers and Paris. In his book Lefebvre pointed out that "all the scientists concerned with Algeria . . . proclaimed unanimously that the conservation of the forests is indispensable to the existence of the colony."[186] He praised the Ligue and noted that its "honorable president, Dr. Trolard, has fought for eighteen years to prevent these devastations."[187]

Indeed, the Ligue had worked diligently throughout the 1890s, especially during Governor-General Cambon's administration. They petitioned the governors-general, local governmental bodies, and the government in Paris. In 1893 they sent a two-hundred-page monograph, *La Question forestière algérienne devant le sénat* (*The Algerian Forest Question before the Senate*), by Dr. Trolard, to all the senators and deputies in the French government.[188] In 1895 they published an essay on "Colonization, a practical program" that highlighted the need for reforestation if colonization were to succeed. In 1897 the Ligue reprinted the Tassy report of 1872, to serve as a reminder and an example of what needed to be done to save Algeria's forests. In 1898 they published a projet de loi for a new Algerian forest code. In short, the Ligue agitated strongly for a new Algerian forest law that would go further than the 1874 and 1885 forest laws had gone, especially in curtailing grazing in the forests and promoting reforestation. The vast majority of its publications drew upon the declensionist environmental narrative to justify their arguments.

Although many believed forest fires to be the worst cause of deforestation in Algeria, some like Trolard believed that pasturing livestock in the forest was the worst cause. Tassy may have been one of the first forestry professionals to raise the cry of alarm when he stated in his influential report that "grazing is the great, the immense, plague of the country."[189]

This would become nearly a mantra for the rest of the colonial period. From the 1870s on, the Forest Service systematically tried to reduce (or eliminate) the livestock of the Algerians on the grounds that "the pastoral life of the natives is incompatible with civilization."[190] Trolard put it succinctly in the projet de loi sent in 1897 to the parliamentary subcommission in Paris studying the question of a new forest law for Algeria: "[G]razing in the forest is . . . the principal cause of the disappearance of the forests; fire is nothing but an accessory cause."[191] In this document he advocated, among other things, confining the Algerians to circumscribed pastures created especially for them away from forests, essentially a form of cantonment. In another of his publications he called for "the fixation to the soil of these three or four million nomads who . . . use and abuse the pastures that still exist."[192] In fact, it seems that Trolard would have preferred to eliminate most of the Algerians altogether, for, as he proclaimed, "Algeria must be populated almost entirely by [the] French."[193] Trolard used the declensionist environmental narrative to justify most of these remarks about nomads and pastoralists in Algeria.

In February 1903, a decade after it was first presented, a comprehensive forest law was finally promulgated for Algeria by the French government. In its 190 articles (based on the 1827 French Forest Code) it retained most of the contents of the 1874 and 1885 laws, including collective punishment and sequestration of property for forest crimes. Many who approved of the new law were quick to point out that it was more flexible and that the fines for infractions were lower than before. In fact, the flexibility allowed for more severe repression and the fines for other than first infractions were higher.[194] The 1903 law was notably harsh on pastoralists. Article 68, for example, denied to the nomads the long-held traditional right of pasturing during summer migration (*achaba*) to the Tell from the south, on the grounds that they deforested. As one author of the period noted, though, this prohibition "had as its objective not only the protection of the forest, but also to impede the movements of the natives from the High Plateaus."[195] On the other hand, the law did leave open the possibility of the government allowing grazing in certain circumstances. It retained the prohibitions of the 1885 law on clearing scrub, or undergrowth, assuming, in the words of a contemporary attorney, that this "scrub was most often nothing but ancient forest, the restoration of which is possible," and that its conservation would be beneficial not only in the mountains but "especially in the regions of the south invaded by the sands of the desert."[196] A postcolonial historian was also of the opinion that those who crafted the law perceived scrub as "nothing but woods ruined by grazing."[197] In this and other ways, the declensionist environmental narrative was written into the 1903 law, as it had been in parts of the 1885 law before it.

Perhaps the biggest innovation in the 1903 law dealt with reforestation. While the 1885 law had allowed for expropriation of land for reforestation that was declared of public utility, especially to protect the water supply, the 1903 law went considerably further. It allowed the governor-general to establish reforestation perimeters, in the name of public utility, inside which owners could not clear land of trees or scrub without the express permission of the Forest Service.[198] These perimeters could be declared for mountain land, for defense against erosion, to protect the water supply, to stabilize sand dunes, for the public health, or to defend territory along frontier zones.[199] In other words, they could be created nearly anywhere in the colony. In addition to the declensionist narrative, then, the 1903 law also incorporated both desiccation theory and the idea that deforestation causes torrents and erosion. Reforestation perimeters began to be declared within a year of the promulgation of the 1903 Forest Code. Following the creation of the Special Reforestation Service (Service Spécial du Reboisement) in 1908, at least nineteen reforestation perimeters comprising 408,000 hectares of land had been declared by the 1930s.[200] By the end of the colonial period, more than two million hectares (of the 7.5 million originally planned) had been placed within reforestation perimeters.[201] The fact that only about 62,000 hectares (3 percent) were actually reforested suggests that the primary objective for the declaration of the perimeters may have been to impede the traditional uses of the land, rather than to reforest it.[202] Indeed, the outcry from the indigenous population was immediate and strong, for their livelihoods had "in certain regions become untenable" due to the 1903 forest law.[203]

The 1903 law further extended the power of the Forest Service in important ways. The law was written to provide a certain amount of flexibility of interpretation, for example in the area of grazing. The question of how it was applied thus became very important. Following particularly severe forest fires in 1902 and 1903, the Algerian government ordered a commission to undertake a new study on "the administration, the conservation, and the exploitation of the forests in Algeria."[204] Although the immediate impetus for forming the commission was criticism of the Forest Service in the face of recent fires, the results in fact governed how the 1903 law would be applied over the next half-century. Ordered by

Governor-General Jonnart in May 1904, the commission presented its findings in early June the same year. It concluded, and the governor-general
agreed, that the Forest Service needed to be strengthened, intensive reforestation implemented, "excessive" grazing suppressed, and indigenous
labor conscripted for forestry projects.[205] The commission's recommendations formed the basis of the six bylaws of August 1904 that completed
the 1903 Forest Code. These and several other subsequent bylaws would
largely determine how the 1903 law was implemented.

Dr. Paulin Trolard, appointed to the commission by Governor-General
Jonnart, was one of the ten reporting members and wrote two of the ten
final reports.[206] These two reports, "Utility of the Forest" and "The Coeffi-
cient of Woodedness," repeated much of what Trolard had been writing
for years, although in a more organized and professional manner. The
commission, with little discussion, unanimously accepted both reports
and their conclusions. One such conclusion was that at least 30 percent of
Algeria should be forested in order to make the country's climate appropriate for European colonization. Since only about 10 percent of the country was then believed to be forested, the commission expressed its opinion that approximately five million hectares had been deforested since
antiquity and urgently needed to be reforested. "Give us good forests, Mr.
Governor [General]," Trolard exclaimed, "and we will give you good
colonization."[207] Trolard thus brought to the commission's work his belief in and particular interpretation of the declensionist environmental
narrative, as did Forest Conservator Lefebvre, also a reporting member.
Trolard considered his appointment to this commission to be one of his
crowning achievements, and boasted that "it officially consecrated all of
our [the Ligue's] ideas."[208] He proudly told his readers in the *Bulletin de
la Ligue* that "the work of this commission was followed by edicts in
which the chief of the administration marked his intentions to execute
them."[209] The chief of the administration, Governor-General Jonnart, was
equally pleased with Dr. Trolard and the Ligue. He joined the Ligue in
March 1905 and in November that year he thanked the Ligue publicly,
saying, "to succeed in this vast enterprise of public interest that is Algerian reforestation, I count a great deal, sirs, on the assistance of the Ligue
du Reboisement, which acquires each day new claims to the gratitude of
the citizens anxious for the prosperity of Algeria."[210]

Despite the triumph of the declensionist environmental narrative
during this period, some critics still raised their voices. When they did,
Trolard or some other member of the Ligue did not hesitate to criticize

their arguments. A couple of years after the 1903 law was passed, for instance, an engineer, H. Dessoliers, wrote a brochure on "Water and Woodedness" ("Eau et boisement") that contradicted much of the Ligue's propaganda. Dessoliers stated that the application of the forest laws "had been absolutely ineffective," and that the amount of available water "had not at all increased since the conquest, and this despite the millions of citations levied against the native, despite the application of collective responsibility, and despite the enormous fines that we inflict on the people after each fire."[211] Trolard devoted several pages in the *Bulletin* to destroying Mr. Dessolier's argument and reinforcing the Ligue's standard propaganda. Critics were not widespread, however, because the narrative was useful to so many different colonial agendas. One area where the declensionist narrative did not penetrate very deeply during this period, though, was the discipline of botany. The work of some Algerian botanists indeed seemed to raise questions concerning much of the narrative so strongly imbedded in Algerian discourse at this time.

A Botanical Counternarrative?

Botanical works written on Algeria before the Third Republic tended to highlight the general fertility of the land, the rich variety of plant life, the beautiful forests, and the abundant pastures. Little of the declensionist environmental narrative is evident in this work.[212] None of these early botanists, however, were residents of Algeria, with the exception of G. Munby, an English colonist who lived near Algiers in the 1840s.[213] Early in the Third Republic, two men who were to have an important impact on botany not just in Algeria but throughout French colonial North Africa moved from France to Algeria. Despite their early membership in the Ligue du Reboisement, over the course of their fifty-year careers the work of these two men, Charles-Louis Trabut (1853–1929) and Jules-Aimé Battandier (1848–1922), seems to have challenged much of the declensionist narrative. Considered the pioneers of North African botany, they published comprehensive works that remained the most authoritative contributions to North African botany for half a century.

Born in France, Trabut arrived in Algiers in 1873 to study medicine. He chose to study in Algeria rather than France so that he could botanize in a new place.[214] He stayed in Algeria upon completion of his medical studies and was named professor of medical natural history at the medical school in Algiers in 1880. Although self-trained in botany, his extensive field experience led to his appointment as the director of the botanical service when it was created by the Algerian government in 1892, a post he held until his death at the age of seventy-six.[215] He met Battandier, who became his lifelong friend and close collaborator in botanical research, in 1876. They lived in neighboring houses for more than forty years, from 1878 until their deaths. Battandier, who was also born in France, obtained a doctorate in pharmacy and was subsequently appointed chief pharmacist for the hospital in Mustapha in 1875.[216] Shortly after this he became professor of pharmacy at the medical school in Algiers. In 1910 he accepted a position teaching pharmacy at the newly established University of Algiers, a position he held until his retirement in 1920. He was killed in an accident at the age of seventy-four. Although he worked as a professor of pharmacy for all of his career, Battandier conducted extensive botanical research, much of it with Trabut. They published many books and articles together over the years, including the major works on Algerian and Tunisian botany from the 1880s into the 1920s. These were the preeminent sources on North African botany until the monumental *Flore de l'Afrique du nord* by their colleague René Maire was published posthumously in 1952.[217]

The vast majority of the research published by Trabut and Battandier was strictly and technically botanical. They produced long lists of species of plants discovered and intricate details of the morphology and physiology of large numbers of plants. It is their more general works, those not strictly botanical, that present an interesting environmental narrative that may be interpreted as a partial counternarrative to the prevailing colonial narrative of the time. In 1891 Trabut authored, with the forester Auguste Mathieu, a special report on the High Plateaus ordered by the Algerian government.[218] The authors began with a description of the soil and climate, followed by a detailed report of the vegetation, its condition, and the prospects for its exploitation. In their discussion of the sparse arborescent vegetation of the region, they described some remnants of ancient woods that had only a few old trees remaining with no new seedlings in evidence. In stark contrast to most other authors of this period, Trabut and Mathieu explained that this situation was caused solely by a gradual decrease in rainfall.[219] They did not blame, as had forester Reynard in his report on the restoration of forests in the south, centuries of overgrazing and abuse by the nomads and their herds of livestock.[220] In fact, Trabut and Mathieu were careful to note that the region with the remnants of old trees had been under the guard of the military and had not been grazed

Figure 4.2. *Right to left:* Louis Trabut, Jules Battandier, René Maire, and Émile Jahandiez in the Experimental Garden at Hamma, Algiers, in 1920. From René Maire, *Les Progrés des connaissances botaniques en Algérie depuis 1830* (Paris: Masson, 1931), 153.

or burned for many years. Yet, despite this, there had been no regeneration of the trees in question.

Later in this report, Mathieu and Trabut considered in some detail the nomadic pastoralists of the region and their traditional livelihoods. They pointed out that these nomadic groups had significant traditional knowledge of livestock raising including the treatment of livestock diseases. The authors argued that, given the environmental conditions under which they lived, the nomads were producing an acceptable number of livestock and that it was incorrect to "attribute, as is generally done, to the ignorance and neglect of the Arabs the weak pastoral yield [head of livestock] in the High Plateaus."[221] They placed no blame on the nomads of the region for environmental destruction. A few pages later, in fact, they identified the French military as the prime cause of what deforestation had occurred in the region, and advised that the military strictly conserve the remaining trees henceforth.[222] They concluded that the environment and ecology of the High Plateaus were most suitable for raising livestock extensively, and secondarily for harvesting alfa grass, the most abundant vegetation.[223] Given the findings of their mission, they suggested several measures for managing this large region. One of the most interesting of these suggestions was that the government should "renounce the constitution of individual [private] property among the natives of the High Plateaus, because a soil appropriate for grazing requires collective use."[224]

Given the ubiquity of the declensionist narrative in Algeria at the time, this report is remarkable for several reasons. First, although it is impossible to tell which author wrote how much or what part of this report, it was very rare for foresters to hold such views at this time. Second, not only did the authors not blame the nomads and their herds for centuries of deforestation and environmental abuse, but they also praised their indigenous knowledge of the environment and of livestock raising. Third, the diminution of trees in the region they explained by a general reduction in rainfall instead of the much more common desiccationist argument that blamed the dry climate on deforestation by the Algerians. Fourth, Trabut and Mathieu criticized the government policy of instituting private land tenure in the region and recommended that it be stopped. This is notable given that Warnier used the declensionist narrative to justify private land tenure in his writings arguing in favor of what became the Warnier Law. The readership of this single report on the High Plateaus, however, was probably not very large, since it was commissioned by the government for a specific purpose. A book published seven years later, however, had similar themes and was widely read in Algeria and in France.

Algeria, Land and People, which appeared in 1898, was coauthored by Trabut and Battandier and published in Paris.[225] The subtitle, *Flora, Fauna, Geology, Anthropology, Agricultural and Economic Resources,* provides an idea of the comprehensive scope of the work. Early in the book, Battandier and Trabut set a tone quite different from that of most works of the period. They stated simply that "until our [the French] arrival in Algeria, the Arab, essentially a pastoralist, had but little disturbed the natural distribution of species."[226] A few pages later they specified, for instance, that the forests of the Tell region were more or less as extensive as they were in ancient times, "even after seven centuries of Arab domination. The Arabs, despite their pastoral mores and their incendiary habits, have left us a magnificent forest domain."[227] They attributed the reputed fertility of the colony in Roman times to better management during that period than during the era of Arab domination, or, for that matter, during French administration. Although they did mention the "Arab invasion," and specifically the Beni Hillal of the eleventh century, they noted that while these "newcomers, being pastoralists and nomads, destroyed the agriculture and all the ancient civilization of the country," they did not destroy the forests or the environment.[228] The authors did, however, condemn what they believed to be current deforestation by the Algerians in some areas due to overgrazing and setting fires in the forest. They attributed part of the problem, though, to overpopulation and the squeezing of the Algerians off the land by European colonization. Later, in their conclusion, Battandier and Trabut were careful to note that in addition to grazing and fire taking their toll on the forests, a slow drying out of the climate was responsible for some reduction in the forested area in Algeria. They explained that, "during the Quaternary, the Sahara seems to have been already a steppe; since then, the steppe has become a desert very slowly."[229] They concluded that the French could restore Algeria to the Roman level of agricultural productivity by careful management of resources, especially of water resources. In the final sentence of their book, Battandier and Trabut stated optimistically that, despite the relatively small climatic changes since Roman domination, "Algeria will be cultivable for many centuries to come."[230] This attitude contrasts sharply with the more common sentiment about the future of Algeria, as expressed by Trolard and others, that the Sahara was menacing the entire country and Algeria was in danger of becoming a desert wasteland.

One of the unique aspects of their book, for the period, is the authors' intimate knowledge of Algerian plant life and the way they related these facts not only to agriculture in the colony but also to the landscape in general. Whereas many partisans of the declensionist narrative described the landscape during the dry season as overgrazed, ruined, and sterilized by the "destructive natives" and their hungry herds, Battandier and Trabut described the adaptations of the local plants to the summer dry season and frequent droughts. They explained that in many regions, including parts of the Tell, "the vegetation stops almost completely during the summer. . . . Yet the herbaceous vegetation, however rare, does not disappear completely. . . . With the first rains, the annual plants germinate en masse, the perennial plants sprout again and the earth is soon covered with a thick carpet of verdure."[231] In their discussion of the Saharan regions, the authors again explained that "the climate alone sterilizes these lands. . . . It rains, however, in the desert, but these rains, sometimes torrential, are extremely irregular."[232] The treeless vegetation of the steppe (the High Plateaus) they likewise attributed to the climate and not to environmental destruction. As in the earlier report written by Trabut and Mathieu on the High Plateaus, Battandier and Trabut noted in this book that the steppes were essentially a region for indigenous pastoralism and for harvesting alfa grass.[233]

Despite the fact that Battandier and Trabut did not condemn the Arabs as harshly as did most colonists at this time, it is important to clarify that they did not romanticize them either. They wrote that "the Arab is neither worse nor better than other men; but their traditional idleness will condemn them sooner or later to disappear before the more active races or to modify themselves profoundly."[234] This paternalistic and bigoted opinion of the Arabs was accompanied by a view of the Berbers influenced by the Kabyle myth. They described the Kabyle favorably as "a worker, economical, intelligent, very suited to business and agriculture, strongly attached to the land."[235]

This book, part of a contemporary science series published by a major Parisian publishing house, was aimed at a French audience. Its goal seems to have been to provide a detailed overview of Algeria, in a positive light. In their preface, Battandier and Trabut remarked that Algeria was still not well known in France and noted that it was often portrayed either as an El Dorado or as a burning hell. Algeria was neither, they wrote, "but it has a little bit of both."[236] While these particular authors, who clearly loved their adopted land, may have been influenced by their effort to portray Algeria favorably, the absence of the declensionist environmental narrative from their writings is mirrored in most of the work of the other Algerian botanists of the period.[237]

In 1888 Trabut presented a paper at the meeting of the French Association for the Advancement of Science (Association Française pour l'Avancement des Sciences), held that year in Oran. The paper was later published in three parts in the *Bulletin de la Ligue du Reboisement.*[238] This essay, "Les Zones botaniques de l'Algérie," was the earliest notable work on the "botanical geography" of Algeria. With its publication, a new kind of botanical study began for North Africa that would soon come to be known as phytogeography.[239] Trabut expanded methodically on the earlier work of the famous botanist Ernest Cosson, who had divided Algeria into four generalized botanical zones: the Tell, the mountains, the High Plateaus, and the Saharan regions.[240] Cosson hypothesized that latitude was the most important determining factor in the distribution of vegetation in Algeria. Trabut, however, pointed out in his essay that the vegetation in each of Cosson's four zones was far from homogeneous and that, in actuality, the distribution of vegetation was determined by a number of different factors, the most important of which were altitude, temperature, and annual precipitation.[241] Trabut did not consider soils or general geology in this essay.

Based on his extensive field experience in most of Algeria, Trabut instead proposed eight botanical zones, each of which he defined and named for the "dominant species" in the zone. Six of the eight had trees as their dominant vegetation. These arboreal zones included the olive, the cork oak, the dwarf palm, the Aleppo pine, the Holm oak, and the cedar. One of the treeless zones was characterized by saline lakes and depressions punctuated with marshes. The other was the zone of the "steppes," which Trabut divided into four subdivisions, one of which he also defined by a tree, *Pistacia atlantica.* The significance of this work lies in how Trabut decided what the dominant species was for each zone, a task on which he placed great importance. He called these dominant species the "characteristic species" and explained that it was those plants that "play a considerable role in the appearance of the country that must be chosen."[242] This method of designating the dominant species, though, was almost entirely subjective, based as it was on the experiences, aesthetic preferences, and judgment of the person evaluating the landscape. Trabut also described the other vegetation of each zone in some detail, in order to further determine each zone's general characteristics. For the zone of the cork oak, for example, he wrote that "all of this zone is rich in forests."[243] By his definition this zone not only contained the cork oak as its dominant species, but it was also essentially a forest zone. He painted a similar picture of the Aleppo pine zone, writing that it was a zone of conifer forests in which the Aleppo pine was dominant. It is significant that Trabut noted that, in several places, the forest had been destroyed, and thus that "one must reconstitute it with the aid of scattered vestiges [of trees] in the maquis."[244]

This last method, of deducing the "normal" vegetation of a region from its "relict" vegetation (usually forests assumed to have been destroyed), was becoming increasingly common in Europe during the late nineteenth century. By the turn of the century, botanists and plant geographers were also considering the influence of soils on the distribution of plants, in addition to altitude, temperature, and precipitation.[245] And although Trabut did not use the declensionist narrative to explain the deforestation he mentioned occasionally in his 1888 essay, later botanists and phytogeographers would. Several of these authors took Trabut's 1888 essay as the starting point for their research. Phytogeographical studies of North Africa began just prior to the First World War. These studies considered the action of humans and animals in deducing the "normal" distribution of plants, in addition to the physical and climatological features outlined above. As the following chapter details, the declensionist environmental narrative was incorporated in these studies, informing the way these researchers deduced what the "normal" vegetation of a region should be, that is, what its "potential vegetation" was. The incorporation of the declensionist narrative in plant ecology would institutionalize the narrative in significant ways that would become perhaps its most important legacy over the course of the twentieth century.

CHAPTER 5

Narrative, Science, Policy, and Practice, 1919 to Independence

BY 1912, THE FRENCH had completed their conquest of the Maghreb with the acquisition of Morocco. As they had done in their occupation of Tunisia in 1881, they carried the colonial declensionist environmental narrative with them to the new protectorate. Unlike Algeria, both new territories were governed as protectorates rather than being directly assimilated to France. The vast majority of policies and legislation applied in the protectorates, though, was adapted from policies and legislation developed in Algeria. Despite some important differences, in Tunisia and Morocco the forests were delimited and managed much as they were in Algeria, property laws were more alike than not in intent and in effect, and the rural populations were squeezed out of their traditional livelihoods in similar ways.

Compared to the conquest of Algeria and Morocco, the occupation of Tunisia was relatively easy for the French. They encountered less resistance in Tunisia and were able to govern with policies that were less draconian than those later instituted in Morocco. The French fought a pitched battle of pacification in Morocco, however, that lasted until the deep south was finally subdued in 1934. Thus, for half of the protectorate period the French were preoccupied with battles of resistance in various parts of Morocco, from the Rif Mountains in the north to the desert regions of the south. In part because of these difficulties in controlling the indigenous populations in the forests and deserts of Morocco, the declensionist environmental narrative proved especially powerful and was incorporated into most environmental legislation and policies in the protectorate.

During the period following World War I the narrative both informed, and was later justified by, developments in plant ecology and the construction of potential vegetation maps, particularly in Morocco. An analysis of these potential vegetation maps and their uses in Morocco provides a case study of how the declensionist colonial environmental narrative regarding North Africa was formalized and institutionalized to become the dominant environmental history of Morocco by the time of independence. Although similar potential vegetation maps were constructed and used in Algeria and Tunisia, the case of Morocco is especially enlightening. Because French administration in the youngest Maghreb colony evolved in step with contemporary developments in plant ecology, the ways in which the narrative influenced environmental policy and environmental history in Morocco are particularly clear. The declensionist environmental narrative was also institutionalized by work in plant ecology in Algeria and Tunisia, and it informed conventional environmental history there as it did in Morocco. Environmental policies in colonial Algeria and Tunisia, though, were already largely set in place, so the impact on them was not as significant as with policies in Morocco. The formalization, quantification, and institutionalization of the declensionist environmental narrative as the environmental history of the Maghreb is one of the most important legacies of the colonial period, because it continues to influence environmental and agricultural policy in the postcolonial states today.

The Consolidation of French Colonial Rule in the Maghreb

French colonial rule in North Africa was consolidated in 1912 with the creation of the Moroccan protectorate, thirty-one years after the institution of the Tunisian protectorate. Despite the substantial resources tapped to successfully occupy and administer these two new territories, French commitment to colonization in Algeria remained strong—so strong that France fought a long and bloody war from the mid-1950s to 1962, when

Algeria won its independence after 132 years of French colonial rule. Part of what fueled the Algerian drive for independence, of course, was the immiseration that resulted from the continuation of colonial policies that disenfranchised the Algerians from their lands and resources and failed to provide adequate employment for those who could not scrape by in the traditional agricultural sector.

The administration in Algeria continued to favor colonization and became increasingly capitalist following World War I. By 1930, the colonists owned about 2,350,000 hectares of the best agricultural land, meaning that roughly 25 percent of all cultivated land in Algeria was owned by settlers, who composed only 2 percent of the total agricultural population.[1] By 1954, settlers and companies controlled 2,726,000 hectares of agricultural land, most of it in large farms or estates. Much of this increase had been facilitated by another property law, passed on 4 August 1926, that subjected the last remnants of collective property to rules very similar to the property laws of the nineteenth century.[2] Between 1928, when the law began to be applied, and 1934, over a million hectares of land were appropriated. From this date land in Algeria was for all intents and purposes fully commodified, and land sales were conducted easily and cheaply. Although the state abandoned official colonization after 1930, property concentration persisted, due to the continued rise of large companies that ran huge agricultural estates for export production. In an ironic turn of events, the settlers themselves were increasingly squeezed out of the agricultural sector by these large companies, which bought up and consolidated agricultural land into massive estates.[3] By the 1930s agriculture in Algeria was largely a capitalist venture conducted by European businessmen, rather than the backbone of settler livelihood, as it had been earlier. Later, by the end of the colonial period, most agricultural revenues were garnered by the export of commercially produced wine, early vegetables, and citrus, almost all produced on such large holdings.

Meanwhile, two-thirds of Algerians, especially in rural areas, were enmeshed in poverty and had little means of escape. Deprived of adequate land to raise crops or livestock by traditional production techniques, many turned to sharecropping and wage labor on the large estates, as they had done increasingly for decades. Statistics vary, but it is likely that between half a million and a million rural Algerians were unemployed in 1954.[4] Out of a total population of about 8.5 million, only 112,000 Algerians had permanent employment in the agricultural sector at the time.[5] Few job opportunities existed elsewhere. In the crisis years of the 1930s, this high unemployment led to a large-scale migration from the countryside to the cities that would continue to the end of the colonial period and beyond. Yet because relatively few jobs were available in the cities, many men (and perhaps a few women) turned to short-term migration to France to find work, sending remittances to their families back home in Algeria. It is estimated that about 15 percent of Algerian men were working in France in 1954.[6] Some efforts were made to ameliorate conditions for the Algerians, but few of these efforts were well-funded enough to make much of a difference. Other measures, like the 1942 Martin Law that was projected to redistribute 15 percent of irrigated land to small farmers, were never implemented.[7] In the face of these miserable conditions, the Algerian independence movement grew in force, and by the mid-1950s a war for independence had begun. In 1962, Algeria emerged from this prolonged and devastating war newly independent from France.

Compared to the difficult, violent, and lengthy battle for conquest and pacification of Algeria, the occupation of Ottoman Tunisia was relatively easy for the French. Using an exaggerated border incident between Tunisia and Algeria as an excuse, the French military occupied Tunisia in 1881. There was surprisingly little resistance to the occupation in most of the country, although many of the nomadic groups in the south did fight for a short time. Rather than implement direct administration, though, as they had done in Algeria, the French set up a protectorate system that maintained the illusion of Tunisian control over the country's affairs.[8] The bey remained head of state, but the French resident general became the de facto ruler of the country. The resident general and his administration controlled all the crucial departments in the government, and, importantly, took over tax collecting and security. Much was borrowed from French experience in Algeria in areas such as forestry, for example. There were important differences, however, in how colonization was managed in the new territory. Free land was not given to French settlers in Tunisia as it had been in Algeria. Land had to be purchased by the settlers, and this resulted in the establishment of a wealthier class of settlers and a smaller settler population.[9] It also resulted in less dislocation of the rural agricultural population, and less expropriation of collective lands. Historians have concluded that this helped to prevent some of the indigenous hostility that was so evident in Algeria.[10]

Nonetheless, 770,000 hectares of agricultural land were owned by French settlers and companies by 1914, representing about one-fifth of the total arable land in the country.[11] By independence, about 850,000 hectares

(22 percent) of agricultural land was in the hands of the French and other Europeans (primarily Italians), mostly in the form of large farms or estates.[12] Of the French land, 23 percent was owned by four large companies.[13] The average French holding was 250 hectares, compared to the average Tunisian's of only six hectares.[14] It has been noted that, due to French policies, on the eve of independence the "dispossessed and proletarianized Tunisian peasants vegetated on overpopulated lands, emigrated to the poorest regions, neglected by colonization, or finally migrated to live in the shanty towns of the large cities."[15]

A further 810,746 hectares of forests had been expropriated by 1889 and were in the hands of the Tunisian Forest Service, which was created in 1883 under the direction of Henri Lefebvre of Algeria.[16] Although Tunisia did not have its own forest law until 1915, it quickly applied some pertinent Algerian statutes. In 1881, for example, the protectorate issued a decree prohibiting the sale of forest land without government permission.[17] But, unlike what they had done in Algeria, the French did not recognize communal forests in Tunisia, and simply defined all forested and potentially forested land as state land at the very beginning of establishment of the protectorate. In 1886, the protectorate passed a forest ordinance that reproduced most of the 1874 Algerian forest law regarding forest fires, including the prohibition of grazing in forests for six years after a fire.[18] In 1888 the government went so far as to decree punishment by death in the case of deliberately set forest fires.[19] These early decrees were later incorporated into the 1915 Tunisian Forest Code, which was amended many times by later decrees, most of which had precedents in Algerian forest law.[20]

As in Algeria, the loss of agricultural and forest land forced many subsistence producers, and especially the pastoral nomads, off their land. Although not as much collective land was expropriated from the Tunisians as had been in Algeria, the alienation of collective lands, including the 140,000 hectares lost between 1920 and 1934 as a result of the 1901 law on the delimitation of collective lands, had very adverse consequences for Tunisia's pastoralists.[21] Much as in Algeria, the Indigenous Affairs Service proudly reported that "one of the results pursued by French colonization in Tunisia has been . . . the fixation of the nomad and semi-nomad tribes to the soil, [and] the progressive and prudent transformation of collective property into individual [private] property."[22] Both the property laws and the forestry laws instated in Tunisia reflected the declensionist environmental narrative that had been developed in Algeria. Indeed, Tunisia as well as Morocco had been incorporated into the declensionist environmental narrative since early in the Algerian occupation. The descriptions of exceptional fertility during the Roman period frequently included all of the Maghreb, as did the later tales of ruin by the hordes of Arab nomads and their voracious herds. Thus the officers of the Tunisian Indigenous Affairs Service could point to "the profound convulsions which, since the Roman era, have upset the country" and assert that "the passage of the Arab armies and later the Hillalian tribal invasion . . . have made of this country a desert strewn with ruins, which, however, attest to its ancient prosperity."[23]

One of the most interesting examples of the use of the narrative in colonial Tunisia involved the massive plantings of olive trees in the region of Sfax at the turn of the century. Paul Bourde, the director of agriculture in Tunisia, used the narrative in 1893 to justify a decree regulating the sale of land in the region that encouraged and facilitated the establishment of olive plantations by European capitalists.[24] This decree, passed in 1892, alienated those (mostly indigenous Tunisians) who did not have legal title or who were occupying too much land according to an earlier law. It also laid out the guidelines for establishing concessions for olive plantations at very cheap rates (ten francs per hectare).[25] Bourde believed that under Roman domination this region "had long ago a great reputation of fertility," and he blamed the "Arab invasion" for its "sterilization."[26] He explained that "before the Arab conquest had deforested and depopulated all of this region, travelers . . . could go from Tebessa to Gafsa always in the shade of forests. . . . Deforestation has produced its work of destruction here as in the rest of the Regency."[27] Reforestation with olive and fruit trees would, according to Bourde, improve the climate and the soil. In 1881, there were about 350,000 olive trees in the region around Sfax. Thanks to Bourde's policies and encouragement, new plantations were begun around the turn of the century, and by independence in 1956 there were more than six million olives and 2.3 million fruit trees in the region.[28] As a result, the landscape was completely transformed. The irregular small farms that had grown a variety of different crops and supported different kinds of livestock were replaced with rigid geometric grids of olive trees nearly as far as the eye could see (see figure 5.1).

It was in Morocco, though, that the French colonial environmental narrative of North Africa reached its apogee. After years of negotiations with other European powers, extensive commercial penetration, and numerous small territorial conquests, the French finally conquered the king-

Figure 5.1. A sea of olive trees, Sfax, Tunisia. The original caption reads, "The 'sea of olives' in the region of Sfax." This photograph shows the extent to which the landscape around Sfax had been changed by the planting of olive trees on a massive scale. From Jean Despois, *La Tunisie* (Paris: Larousse, 1930), 157. Reproduced with the kind permission of Librarie Larousse.

dom and established the protectorate of Morocco in 1912.[29] From the first years of the protectorate, the myth of the granary of Rome drove much of Morocco's colonial agricultural policy, especially the early development of the cereals sector. The first resident general of Morocco, Louis H.-G. Lyautey (1854–1934) himself proclaimed that Morocco had been one of the granaries of Rome.[30] In the late 1920s a different classical fertility narrative emerged, that of the lush garden of the Hesperides, which had been located in Morocco by classical authors.[31] It was used to stimulate and to justify what became the massive citrus and irrigated vegetable production in the humid northwestern coastal plains that still dominates agriculture and agricultural policy in Morocco today.[32] In policies and legislation concerning property and forestry, though, the complete declensionist environmental narrative was invoked. Most laws and policies were adapted from those already in use in Algeria, and some from those used in Tunisia. Lyautey, who came to Morocco with many years of experience in Algeria, Madagascar, and other French colonial territories, strove to avoid what he perceived as the pitfalls of previous French colonial administrations.

Despite these intentions, he implemented a land registration act in 1913 that was in many ways similar to those in Algeria and Tunisia. This law was designed to facilitate the appropriation of the best agricultural lands by French colonists, and by 1932 some 840,000 hectares were in French hands.[33]

In an effort to protect the collective lands that were the domain of the nomadic and seminomadic pastoralists, Lyautey passed a law in 1914 that declared collective lands inalienable. This law was reformed by a 1919 law that placed collective lands under the guardianship of the state and governed their use.[34] It allowed collective lands to be leased both for the long term and in perpetuity, provided for their expropriation in the interest of "public utility," and let the state create colonization perimeters.[35] It was after this law was passed, in effect enacting a similar policy to that of cantonment in nineteenth-century Algeria, that the largest colonial acquisitions of Moroccan lands were made. Furthermore, all forest land had been claimed by the state at the beginning of the protectorate. As in Tunisia, the Moroccan administration did not even attempt to delimit "communal forests." Forest lands were delimited and protected for the "public good" and for the restricted use of a few "local tribes." Thus a further 3.8 million hectares of forest land were lost to many of the Moroccans who had traditionally used them by 1916.[36] By the end of the protectorate period (1956) over a million hectares of the best agricultural land were controlled by French colonists, only 40 percent of the arable land was owned by Moroccan smallholders (most had only a few hectares), and 3.6 million rural Moroccans were landless.[37] Deprived of land and resources and taxed heavily, rural Moroccans were pushed toward big cities like Casablanca and their growing shantytowns (*bidonvilles*). This rural exodus became a flood in the 1930s, and rural out-migration has remained a problem into the postcolonial period.

In sharp contrast to their easy experience occupying Tunisia, the French encountered pronounced difficulties in conquering Morocco. From the inception of the protectorate, the French fought Moroccan peasants in the mountains of the northern part of the country and pastoral nomads in the south.[38] This battle of pacification lasted for half of the colonial period, until 1934, when the pastoralists in the deep south of the country were finally subdued. Because of the political troubles engendered by the "unruly tribes," the protectorate administration was preoccupied for much of the colonial period, even after final pacification, with controlling and attempting to sedentarize them. With their property laws, forestry laws, and administrative policies, the French in effect attempted "a strategic restructuring of the tribe inspired by a mission to discipline and punish."[39] Thus, many Moroccan policies were more harsh than those implemented in Tunisia, including the Forest Code, which was derived nearly entirely from the Algerian code and was much more comprehensive than the Tunisian code. Likewise, the difficulties the French encountered with nomadic and seminomadic pastoralists in Morocco appear to have led to policies dealing with grazing and pastoral migration of a severity not encountered in Tunisia. Broadly speaking, nearly all environmental policy in colonial Morocco, outside of urban and agricultural areas, was developed or strongly influenced by forestry and secondarily by what environmental scientists call today range management. Both of these sectors incorporated the declensionist environmental narrative of North Africa in their legislation and applied policies.

Morocco had been an integral part of this colonial environmental narrative from its inception, as Tunisia had been. One typical book on North Africa, written just a few years after the capture of Algeria, reiterated the ancient Greek geographer Strabo's observation that "all of the [land] between Carthage and the Pillars of Hercules is of an extreme fertility."[40] In today's terms, this area of assumed extreme fertility stretched from contemporary Tunisia to Morocco's Atlantic coast. From at least the mid-nineteenth century on, Morocco was often singled out as, in the words of an 1854 text, "one of the most beautiful and fertile countries of the earth."[41] It, too, was frequently described as "one of the granaries of Rome."[42] In the decade or so leading up to its conquest in 1912, such descriptions of Morocco's ancient fertility and lush vegetation became more common and more detailed.

Although little actually was known about Morocco at the turn of the century, lengthy descriptions of Roman Morocco as far south as the Draa River on the Atlantic coast extolled its marvelous fertility. One author in particular, Maurice Besnier, depicted the territory's rich resources in authoritative and extensive tracts for the Scientific Mission of Morocco (Mission Scientifique du Maroc), a research group commissioned by the French government. Based on his readings of the ancient texts of Strabo, Pliny, Pomponius Mela, and others, he proclaimed that in the past, "except for a few deserts of small extent, Mauretania [Morocco] [had] only fertile land and [was] well supplied with streams. It [was] very forested, and the trees [attained] there a prodigious height."[43] This environmental narrative became enshrined in *Les Archives marocaines*, a series of volumes produced by the Mission Scientifique du Maroc from approximately

1903 into the 1930s. As in Algeria and Tunisia, the Moroccan narrative primarily blamed the Arab nomad invasion of the eleventh century for the general destruction of the formerly lush landscape, and especially for assumed deforestation. The influential geographer Jean Célérier, for instance, taught the young recruits in the Indigenous Affairs Service that the "thieving nomads," the "undisciplined hordes" who invaded North Africa in the eleventh century had destroyed the region. Quoting Ibn Khaldoun, he concluded that "every country occupied by the Arabs is a country ruined."[44] This culpability was extended to the Arab pastoral nomad populations living in the region since that time. It was frequently deplored, from the early days of the protectorate, that in the Moroccan countryside, "as everywhere, the Arab has destroyed the tree."[45] Such assumptions about the generally degraded state of Morocco's forests provided an imperative to resurrect the forest in Morocco from the earliest days of French control. The declensionist colonial environmental narrative of North Africa thus underpinned the creation, development, and work of the Moroccan Forest Service.

Resurrecting the Forest: The Development of Forestry in Morocco

In 1954, two years before independence, Paul Louis Jules Boudy (1874–1957), the honorary inspector general of the Moroccan Forest Service, presented a triumphant historical overview of four decades of the service's work. Titled "La Résurrection de la forêt marocaine," it began, "In 1913, the day after the establishment of the protectorate of France, the Moroccan forest was in a miserable state and its trajectory tended towards absolute zero."[46] This quote succinctly summarizes the views of Boudy, the man who created the Moroccan Forest Service, directed it for three decades, and maintained strong influence in the department until Moroccan independence. Boudy's views of North African and Moroccan environmental history mirrored the predominant views of the period and are clearly reflected in how he shaped the Moroccan Forest Code and enacted forestry policies. He believed passionately that Morocco had enjoyed extensive forests in the past and that, if destructive local forest use could be prevented, the protectorate had great potential for extensive forests in the future. A forester by training, he rose to positions of unusual power in Morocco within and outside of the Forest Service.[47] Indeed, Boudy's importance and illustrious career have been noted by many, including the "father" of forest ecology, Philibert Guinier. Guinier wrote warmly of

Boudy's prestigious career as a great forester and was of the opinion that he had played "a role perhaps unique in the history of global forestry."[48]

Paul Boudy was born in France and graduated from the National Forestry School at Nancy in 1897 at the age of twenty-three (see figure 5.2). In 1898 he was posted to Algeria, and then to Tunisia in 1904. In 1907 he joined the reforestation department back in Algeria, of which he later became director.[49] In March 1912, the beginning of the protectorate, Boudy volunteered to go to Morocco to help with forestry in the new territory. In March 1913 he was appointed director of the protectorate's newly established Forest Service. Boudy ran the Forest Service for nearly thirty years, until 1941, then stayed on as an official adviser to the protectorate administration until his death the year after independence.

Resident General Lyautey ordered the creation of the Forest Service (Service des Eaux et Forêts) in 1913, and it was placed under the general direction of the Department of Agriculture, Commerce, and Colonization. The first job of the new Forest Service, by order of Lyautey, was to save the cork forests along the mid-Atlantic coast (Mamora).[50] This interest in

Figure 5.2. Paul Boudy. Boudy, pictured here with his 1897 National Forestry School graduating class, is second from the right, wearing a large beret. From Archives de l'École Nationale du Génie Rural des Eaux et des Forêts (ENGREF), Nancy, France. Reproduced with the kind permission of ENGREF.

cork followed a 1912 cork harvest in Algeria that was the largest and most lucrative to that date.[51] The Moroccan cork forests were considered degraded from indigenous abuse, but capable of being saved, reconstituted, and made similarly profitable. As in Algeria, the environmental narrative of ruin served the interests of some groups in Morocco much more than others. The protectorate administration, for example, declared ownership over all forested areas largely in the name of environmental protection, but, importantly, it also generated substantial revenue from the sale of cork, timber, and other forest products that was used to finance French administration of the territory. Revenue from forest production grew quickly during the first few years of the protectorate. Between 1915 and 1918, for instance, it increased over 800 percent.[52]

When Boudy arrived in Morocco, a veteran of twelve years in the forestry departments of Algeria and Tunisia, he likely already had an idea of what he would find in Moroccan forests. In 1914, he published an article on the forests of Morocco that made clear his opinion that the known forests in the protectorate at that time, especially the cork forests, were badly degraded and deforested.[53] Fire, charcoal making, debarking for tannin production, and grazing—that is, most of the indigenous uses of the forests—all were blamed for forest degradation, as they had been in Algeria and Tunisia. At that time, just two years into protectorate administration, only small areas of northwest and eastern Morocco had been explored and their forests inventoried. It was not until well after World War I that the pacification of Morocco was completed, and exploring and inventorying Morocco's forests likewise took decades. Boudy's predecessors, Forest Inspectors Dupont and Lapie, had been sent from Algeria in 1911 to evaluate the forests of Morocco and to write an advisory report. Based partly on their report, the protectorate government decided in 1912 that the Algerian Forest Code of 1903 should be applied, though adapted to Morocco's conditions.[54] The Algerian code was the working forest code in Morocco for five years and was the basis for the new Moroccan Forest Code of 1917.

Resident General Lyautey wanted to avoid in Morocco what he considered the errors made in Algeria and Tunisia, especially where protectorate interaction with the indigenous population was concerned. One of his goals was to pacify by attraction and not with the coercion and violence that had occurred so often in Algeria and elsewhere.[55] This meant, in theory, that local custom and use rights were to be more respected and that application of the Forest Code was to be less harsh than in Algeria.

In setting up the Forest Service then, Lyautey determined that, "contrary to what happened in Algeria, the Moroccan forestry personnel [would have], in effect, a role exclusively technical."[56] At this early date, infractions were to be dealt with by the local authorities and not forestry personnel. Most of the changes made to the Algerian Forest Code as it was adapted to Morocco thus related to how and where to apply the code and did not modify any of the code's primary goals or ecological bases. Whereas in Algeria the French had carried out a lengthy and difficult process of defining state forests, communal forests, and private forests, in Morocco the administration simply appropriated all forested land and defined it as state land to be reserved for the public good and the restricted use of a few local groups. This was mandated with a government circular in November 1912, which became law on 7 July 1914. It was followed by the law of 3 January 1916 on auditing and delimiting state forests.[57]

The Moroccan Forest Code, conceived by Boudy and formalized in October 1917, was based on the Algerian Forest Code, although it was shorter and had changes in modes and extent of application. This legislation itself carried the conventional French colonial vision of the North African environment to Morocco in many ways, since the narrative was embedded in parts of the Algerian Forest Code. Although the two codes were similar, the Moroccan code contained only 84 articles, compared to 190 articles in Algeria's Forest Code.[58] Most of the fines for infractions of Moroccan forest laws were lower than those in Algeria. The 1917 forest law (*dahir*) allowed limited use of the forests by Moroccans only, and not by European colonists, which was a significant change from the Algerian code. The Moroccan code severely restricted the use of fire in or near forests by the few "local tribes" allowed to use portions of the forest, and completely banned it elsewhere. Charcoal production and the collection of traditional forest products such as firewood, foods, and medicinal plants were likewise severely restricted and regulated. As in Algeria, grazing was expressly prohibited for six years in forests that had burned or been harvested for cork or timber. Most traditional uses of the forests, which had previously sustained entire communities, were restricted, regulated, or criminalized.

Due to the slow progression of pacification in Morocco and the disruptions of the First World War, the new Forest Code was not widely applied until the 1920s, and more extensively after 1935 with the final pacification of the south. In fact, the Moroccan Forest Code contained a provision with no precedent in any form in Algeria or Tunisia. This was article two, which stated that the Forest Code would be applied to different territories successively, by governmental decree, as they were conquered.[59] Areas outside of these decreed territories were covered by separate government rules.[60] As pacification progressed in Morocco, more and more of the forested areas were explored, inventoried, and eventually brought under the rules and regulations of the 1917 Forest Code. The initial estimates of about two million hectares of forests in the protectorate were revised during this period of expansion. At the end of the protectorate period, the forested areas were said to cover approximately five million hectares.[61] Estimates of deforestation also changed over this period, in extent as well as in modes of calculation. Tracing these changes reveals an interesting history of policy formation at the same time that it illuminates a significant, but overlooked, period in the history of plant ecology. During these formative years of the 1920s and 1930s, as the new science of plant ecology was being developed, the narrative of environmental decline and deforestation was refined, quantified, and institutionalized in Morocco by Boudy and others working in the protectorate. What had been during the nineteenth century rather vague generalizations about deforestation, overgrazing, and environmental degradation came to be formulated as exact, scientific facts. This Moroccan research conducted in the 1920s and 1930s drew heavily on earlier work in botanical geography, also called phytogeography, in France as well as in Algeria. These historical precedents had an important impact on phytogeographical studies in the new protectorate.

French Phytogeography and Its Application in North Africa

The two men considered to be the founding fathers of botanical geography both worked extensively in France but were not French by birth. One, Alexander von Humboldt (1769–1859), was German but he spent much of his career in Paris. The other, Augustin Pyramus de Candolle (1778–1841), was Swiss and he also lived in Paris for a substantial part of his career.[62] Humboldt made many contributions to plant geography and is famous for his research that correlated the distribution of vegetation types with factors such as elevation, latitude, precipitation, and temperature.[63] He is credited with being one of the earliest scientists to advocate defining a plant association (or zone) by its one, or perhaps two, "characteristic" species.[64] His essay on "plant geography" was published as a sort of appendix to de Candolle's essay on botanical geography in 1820.[65] De Candolle, a friend of Humboldt's, also used this methodology and his essay on botanical geography was one of the most influential in the nineteenth century. Phytogeographical research was presented in greater detail by de Candolle's son, Alphonse, in his important and frequently cited *Géographie botanique raisonnée* of 1855.[66] Although Alphonse de Candolle introduced a more statistical methodology to botanical geography, the primary methods of identifying the dominant, or characteristic, species of each zone remained quite subjective, relying on the botanist's expertise. Thus, "in effect, the characteristic species [was] not necessarily the most abundant or the most visible, it could even be lacking in certain groupings."[67]

Charles Flahault (1852–1935), a professor of botany at the University of Montpellier, was a militant proponent of phytogeography and of defining vegetation associations by their dominant species. He was an important figure in the history of French phytogeography, in which he is considered a pioneer, and he had a substantial following, sometimes called the Flahault school.[68] Flahault worked on French vegetation and especially French forests, which, given the existing soil and climatic conditions, he believed were badly degraded. He frequently identified the dominant species of his vegetation associations by the "relicts" of what he believed had been

Figure 5.3. Charles Flahault. Reproduced with the kind permission of the Hunt Institute for Botanical Documentation, Carnegie Mellon University, Pittsburgh, Pennsylvania.

the natural vegetation before it had been degraded. He wrote in 1896, for example, that "the actual state of the vegetation in our civilized countries no longer represents the primitive [natural] state. . . .The primitive vegetation has disappeared; in all our mountains, abusive exploitation of wood has modified profoundly the composition and distribution [of the vegetation]."[69] Researchers needed, in his opinion, to find "some of the normal elements of the primitive associations; thanks to them we can reconstitute [the natural vegetation]."[70] Flahault was one of the earliest phytogeographers to combine the method of using relict vegetation to deduce the "natural" or "primitive" vegetation with the method of defining vegetation zones by their dominant species. In this way, based on existing soil, climatic conditions, altitude, and so on, phytogeographers believed they could reconstruct what the natural vegetation *should* be (or could potentially be) by using plants, usually trees, that they assumed were remnants of an earlier vegetative cover.

Sometimes such assumptions were accurate and sometimes they were more fanciful than true. This general approach to deducing what the "natural" vegetation should be (the potential vegetation) has informed plant geography and ecology in general to the present, especially in the form of Braun-Blanquet's legacy, the "relevé method," which relies on the scientist's selection of what represents the presumed natural vegetation or ecological dominant. This method, although criticized over the years for being too subjective, remained the most widely used outside of North America until the late twentieth century. Since the 1990s, it has been increasingly common in the United States as well.[71]

The botanical geographical method of identifying vegetation zones or regions and defining them by their "dominant" species also began to be applied to North Africa during the late nineteenth century. It was evident in the work of Louis Trabut, especially in his 1889 article "Les Zones botaniques de l'Algérie," discussed above. In this article, Trabut identified eight botanical zones and defined each by its dominant species. He relied, though, as did Flahault in France, on what he considered relict trees to reconstitute the "natural vegetation" in some parts of Algeria that he believed to be deforested. Trabut appears to have been the first botanist working in Algeria to employ the relict method of deducing the "normal" vegetation and dominant species of the botanical regions of the country. Trabut and Flahault were both active members of the Société Botanique de France, which met several times in Algeria in the late nineteenth and early twentieth centuries. They both published regularly in the society's

Bulletin and read and cited each other's work. They may have known each other well, since Flahault's brother moved to Algeria in the 1870s and Flahault may have visited him and met his fellow botanists Trabut and Battandier.[72] They certainly met when they attended the conference of the Société Botanique de France in 1906, held in Oran. Battandier also attended this conference, as did a young botanist named René Maire who would become a key link between Algerian and Moroccan botany and ecology.[73]

Born in France, René Maire (1878–1949) obtained a doctorate in the natural sciences in 1902 in Paris and, in 1916, a doctorate in medicine at Algiers.[74] He taught in Nancy from 1902 to 1908 and accepted a position in 1911 as the chair of botany at the University of Algiers, where he remained for the rest of his long career. In 1930 Maire was appointed the director of the Botanical Service of Algeria, succeeding Trabut in this position. Maire met and established a lifelong collaborative friendship with Trabut and Battandier during the 1906 conference. From the time Maire arrived in Algiers in 1911, he botanized regularly with Trabut and Battandier in many different parts of Algeria. Maire also published several botanical works with Trabut and Battandier, including the *Atlas de la flore de l'Algérie* in 1920 and the influential *Carte phytogéographique de l'Algérie-Tunisie* in 1925.[75] In addition to his other duties, Maire was charged with exploratory botanical missions in Morocco from 1921 to 1930 by the Moroccan administration. During this decade he made twenty trips to study the botanical treasures of the new protectorate. Maire met and became a mentor to the young botanist Louis Emberger when the latter arrived in Morocco in 1923. Over the next several years Emberger became Maire's principal collaborator and they published over a dozen scientific articles and books together.[76] Although officially based in Algeria throughout his career, Maire published more on Moroccan vegetation than on Algerian vegetation, and he maintained an active research interest in Morocco until his death in Algiers.

One of Maire's most notable publications on Morocco was a chapter in a 1921 government publication on the vegetation of Morocco that provided an overview of most of the protectorate under French control at that time.[77] In this chapter he divided Morocco into fourteen botanical regions, nearly all defined by their arboreal vegetation or their potential arboreal vegetation. He used the relict method to deduce what the "normal" or potential vegetation should be for most of these regions. The two regions of dunes were the only ones not defined by relict vegetation or described as degraded. Nine of the fourteen regions were described as degraded by agriculture or by grazing and burning and were defined by what Maire considered their relict arboreal vegetation. Regarding the steppes of eastern Morocco, for instance, Maire wrote that "the spontaneous vegetation was, based on the few relicts that have escaped grazing and cultivation, a forest-parc [savanna] of Betoum (*Pistacia atlantica*) and Jujubier (*Zizyphus lotus*)."[78] Here and throughout this chapter Maire clearly uses the phyto-geographical method of identifying vegetation zones and defining them by what he believed was their dominant species—whether they currently existed or not. Maire advocated studying religious sanctuaries (*marabouts*) in Morocco to help determine original, "natural" vegetation (see figure 5.4).[79] He saw such areas as small islands of relict vegetation that could point the way toward the reconstruction of the potential vegetation of the region in which they were located.

In the Middle Atlas region, Maire decided that the "natural" vegetation was a forest dominated by evergreen oaks (*Quercus ilex*) and junipers. He wrote that this original forest had been "destroyed by fire and intensive grazing and nothing remains but some scrub . . . and some rocky pastures."[80] He deduced this from the "rare grazed bushes of *Quercus ilex* that still attest to the ancient existence of the forest."[81] As demon-

Figure 5.4. The sacred grove at the tomb of the sultan, Tlemcen, Algeria. Religious sanctuaries such as this were often sought out by botanists for the vegetation that grew in and around them. Such vegetation was usually considered to be "natural," despite the fact that such holy sites were often built near springs and intensively cared for by humans. From Pierre Dumas, *L'Algérie* (Grenoble: B. Arthaud, 1931), 148.

strated above, however, pollen core analysis indicates that in this roughly 50-by-100 kilometer region the presence or number of evergreen oaks does not appear to have changed significantly over the last four or five thousand years.[82] Neither do these data show the extensive deforestation that Maire presumed in this and other regions of Morocco. The relict method of deducing the "natural" vegetation was incorrect in this case as it was in many others. Maire highlighted in this article, and many of his other publications, what he believed to be the destructive action of humans and livestock on vegetation, and he particularly condemned grazing, fire, and agriculture. He did not invoke the full declensionist environmental narrative, though, and he did not blame any particular humans, Moroccans or otherwise, for the degradation.[83] Nor did he attempt to numerically quantify deforestation in the protectorate. He did, however, indicate that Morocco had suffered significant deforestation in six of the nine "naturally" forested regions. It was Maire's colleague and collaborator Louis Emberger, working with Paul Boudy, who quantified Moroccan deforestation and, in the process, fully invoked the declensionist colonial environmental narrative.

Mapping the Potential: French Plant Ecology and North African Deforestation

One of the earliest attempts to quantify Moroccan deforestation was made by forestry director Boudy in a 1927 lecture to the new recruits of the Indigenous Affairs Service (Service des Affaires Indigènes). This department was tightly allied to the administration and performed a variety of intelligence services for the protectorate. Conferences were held annually for new recruits and the lectures served not only as an introduction to the service but also as indoctrination in correct ways of thinking and acting. Boudy began his lecture by asserting the crucial role that forests played in the climate and economy of a country, emphasizing that a "certain proportion of wood is in effect indispensable for a country to actually be inhabitable."[84] The lecture contained numerous such generalizations, which were very similar to those Trolard and many others had made in Algeria and Tunisia.[85] One idea that Boudy particularly highlighted throughout his career was that pastoralists, especially nomadic pastoralists, destroyed forests and caused desertification. As he explained to the new recruits, "there exists, in effect, a very tight link between Semitic [Arab, not Berber] nomadism and vegetal devastation."[86] Boudy adhered to the "Kabyle myth" often used in conjunction with the declensionist environmental narrative in Algeria and Tunisia, and he explained that "the Berbers have respected their woods much better [than the Arab Semites] and they have not destroyed them except for military necessity."[87] Elsewhere in the lecture he proclaimed that "it is scientifically proven that the formation of sand deserts . . . is due to deforestation, the work of ancient nomadic peoples who resort to fire to procure pasture for their beasts."[88]

Following this impassioned introduction, Boudy explained that to be able to support "social organization," any given country must have 30 percent forest cover. This ratio is referred to as the ratio of woodedness, or the *taux de boisement*. He then calculated that Morocco had a ratio of woodedness of only about 9 or 10 percent, based on his estimates of forest cover. Since he believed Morocco should have a rate of woodedness of 30 percent, and must have had that in the past, as Roman civilization had apparently flourished in Morocco, he assumed that the country was approximately two-thirds deforested. He blamed the deduced deforestation on burning, grazing, and clearing land for agriculture by the indigenous Moroccans over the previous centuries. If this destruction could be controlled, Boudy believed, Morocco had the potential to have at least six million hectares of forest, three times the amount believed to exist at that time.[89]

By 1934, the year of the final pacification of the south, forest cover in the protectorate was estimated to be about three million hectares. During the intervening seven years, Boudy had revised his estimate of deforestation, claiming nearly 85 percent deforestation across the protectorate.[90] He calculated this amount of deforestation in an article on Moroccan forests that he coauthored with botanist and ecologist Louis Emberger for a special volume, *La Science au Maroc,* written for the fifty-eighth meeting of the French Association for the Advancement of Science, held in Morocco that year. This estimate of deforestation was detailed further in an article by Emberger in the same volume on general Moroccan vegetation. Emberger estimated that seventeen million hectares of forest had been destroyed. He calculated this based on the three million existing forested hectares and the presumed twenty million hectares that he thought should be naturally forested.[91] He emphasized that at least four million hectares of this deforested land needed to be reforested for ecological equilibrium to be restored. As Emberger described it, "the traveler who traverses Morocco following the classical paths [itineraries described by ancient writers] comprehends with difficulty the profound physiognomy of Moroccan vegetation. . . . Nothing but ruins! Man aided by flocks has made war on the forest and brutally exploited it for centuries. Immense areas formerly covered are today bare."[92] He attributed this massive amount of deforestation (85 percent), as did Boudy, to centuries of abuse by Moroccans and their flocks. Boudy worked with Emberger for many years and was deeply influenced by his research.

Louis Emberger (1897–1969) was born in France. He obtained his doctorate in the sciences with a dissertation on plant cytology in 1921, having first studied to be a pharmacist. He took a post teaching at the pharmacy school in Montpellier in 1921.[93] It was here that Emberger met Flahault, who initiated him in the methods of phytogeography. Emberger developed a strong and intimate relationship with Flahault over the next several years and later married Flahault's daughter. In 1923 Emberger went to Morocco for the first time and was "seduced by the country."[94] He decided to give up his teaching position to move to Morocco. In 1926 he was appointed director of botany at the Moroccan Scientific Institute (Institut Scientifique Chérifien) and professor at the Moroccan Institute of Advanced Studies (Institut des Hautes Études Marocaines) in Rabat. Emberger remained in Morocco until 1936, conducted a prodigious amount of research, and quickly became the preeminent botanist/ecologist in the protectorate. In 1936 he returned to France, and in 1937 was appointed the chair of botany at the University of Montpellier, where he later founded the prestigious "new" Institute of Botany in 1959 and the influential Centre d'Études Phytosociologiques et Écologiques (CEPE) in 1961.[95] Emberger maintained a strong interest in Morocco and North Africa throughout his career and published on plant ecology in the region until his death at the age of seventy-two.

Emberger's research on Moroccan vegetation coincided with a period of great activity and advancement in the young discipline of plant ecology. Many influential and formative works in plant ecology were written from the 1890s through the 1930s. Concepts of plant associations and formations were honed, many of them based on mapping rainfall and temperature region by region, following the work of Humboldt, the de Candolles, Wladimir Koppen, and Johannes Warming. Phytosociology (French plant ecology) was growing in prominence in France with the work of Josias Braun-Blanquet and others. Clementsian succession and ideas of climax vegetation were gaining ground in American ecological studies (although there were critics of Frederick Clement's theory, such as Henry Gleason). Arthur Tansley's ecosystem concepts were being refined in the

Figure 5.5. Louis Emberger. This photograph shows Emberger in 1966 at age sixty-nine. Reproduced with the kind permission of the Hunt Institute for Botanical Documentation, Carnegie Mellon University, Pittsburgh, Pennsylvania.

United Kingdom and subsequently enjoyed widespread acceptance. It has in fact been said that "the first quarter of this century was the age of the great general theories of vegetation."[96] Emberger embraced many of these new developments and worked to contribute to the new science of plant ecology from his base in phytogeography. Emberger's contribution to these general vegetation theories has been mostly overlooked except in studies of Mediterranean ecology. Here he is nearly universally credited with defining and delimiting Mediterranean climate and vegetation zones, called bioclimatic zones.[97] His pluviometric quotient, which takes into account the effects of temperature, rainfall, and evaporation on plant associations, has been especially influential.

Two elements of Emberger's work are not widely appreciated but are important for both environmental history and environmental policy in Morocco and more widely in the Mediterranean basin. The first is that Emberger defined the Mediterranean bioclimatic zones (*étages*) on the example of Morocco, which he believed "alone possesses the complete series of these stages [zones]."[98] The second is that Emberger defined all of his five zones in Morocco by their trees (or potential trees), based on his belief in the declensionist narrative (see figure 5.6). He believed that even the most arid zone in the Mediterranean basin (*l'étage méditerranéen aride*) naturally had forest vegetation, though "the forests there are usually very thin and comparable to savannas."[99] He explained in his defini-

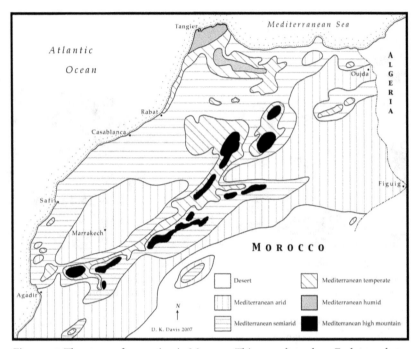

Figure 5.6. The stages of vegetation in Morocco. This map shows how Emberger drew the main vegetation zones of Morocco as he conceived them in about 1934. Five of the six vegetation zones depicted here (excluding the desert zone) are defined by their trees or potential trees. This map is a precursor to the much more detailed phytogeographic map of Morocco that Emberger created in about 1938. Modified from Louis Emberger, "Aperçu général [de la végétation]," in *La Science au Maroc*, edited by Paul Boudy and P. Despujols (Casablanca: Imprimeries Réunies, 1934), 152.

tive publication on Moroccan vegetation that his work had "been oriented toward researching the primitive [natural or potential] state of the vegetation, the climax."[100] Following Maire's example, Emberger looked in places like religious sanctuaries (marabouts) for this relict vegetation, as American geographer Marvin Mikesell would do a generation later.[101] Although such sanctuaries are often intensively cared for by humans and not infrequently built near springs, the vegetation in and around such sites was assumed to be natural. This is how Emberger was able to estimate that twenty million hectares in Morocco naturally should be wooded; he inferred it from relict vegetation and bioclimatic maps. Such deductions from relict vegetation, though, were and are fraught with serious problems.[102]

Much has been learned in the field of paleoecology during the last several decades. Research in this field, especially fossil pollen analysis, points

Figure 5.7. Imilchil, Morocco. The original caption to this undated photo from the protectorate of Morocco reads, "In an arid and denuded landscape." Such images and captions helped to reinforce the degradation story throughout the protectorate period in both popular and official publications. This village in the High Atlas Mountains is, however, located in a rain shadow, where it receives very little precipitation and thus "naturally" lacks significant vegetation. From Centre des Archives Diplomatiques de Nantes, France, "Protectorat Français au Maroc: Fonds Iconographique." Reproduced with the kind permission of the Archives du Ministère des Affaires Étrangères, Nantes, France. Photograph by Jacques Belin.

to some deforestation in a few areas of Morocco, but not in many others.[103] This body of research contradicts Emberger's calculations, for it documents no definitive overall pattern of massive deforestation on the order of 66 to 85 percent over the last two millennia. Moreover, plant ecology has recognized disturbances—fire, disease, and erosion, for example—as integral and natural parts of the ecosystem.[104] From this point of view, a single ideal or climax vegetation is not the natural culmination for plant succession, because multiple stable states with very different vegetation are possible, depending on disturbance regimes. Thus grasses, not trees, may form the "climax" vegetation in many arid regions and regions with frequent fires or heavy grazing. This new research calls into question the very concept of the potential ("natural") vegetation map. Experimental succession studies and paleoecological research, though, were extremely limited in the early twentieth century. It was common, therefore, to use relict vegetation, assumed to indicate the remains of natural vegetation, to reconstruct both past vegetation and potential, or natural, vegetation—

in other words, what the vegetation *could and should* be. This approach, as outlined above, had a strong historical precedent in France and parts of Europe. Emberger was introduced to these methods first by his father-in-law Flahault and soon thereafter by his colleague and mentor Maire.

Emberger explained in his work that it is possible "to deduce the primitive [natural] state of vegetation of a site today bare, but where the ecological conditions and, particularly, climate are the same as in the locality with intact vegetation."[105] Emberger believed, as did Maire, that in Morocco "the actual vegetation cover, over immense surfaces, represents degraded states, the 'miserable remains' of a past much richer and above all more wooded."[106] He blamed the ruined state of vegetation in the protectorate on the Moroccans and their herds who had made "war on the forest and exploited it brutally for centuries."[107] This led to authoritative conclusions, such as that "the absence of red juniper on the south slope of the Middle Atlas cannot be due to anything but the complete destruction of this tree."[108]

The colonial environmental narrative was important in shaping much of Emberger's research, especially his phytogeographic map of Morocco prepared in 1938. This map is still frequently referenced today as the definitive vegetation and bioclimatic map of Morocco. Given the climatic conditions, soils, and elevation of any given region, this map shows Emberger's deductions as to what the natural or potential vegetation should be, based on subjective interpretations of existing and relict (and sometimes absent) vegetation. Of the twenty vegetation zones he identified, seventeen are defined and named for what he considered the dominant tree species, whether they were the most common plants in the zone or not.[109] Often, these "dominant" trees were what he believed to be the "miserable remains" of previously large forests. He considered the arboreal vegetation in nearly all of these zones to be badly deforested and he drew special punctuated lines on this map to indicate areas especially degraded. Emberger's work did more than incorporate the ubiquitous declensionist narrative, though; his research and writings refined, quantified, and firmly institutionalized the declensionist narrative in the science of plant ecology in Morocco. His work, in particular his 1938 phytogeographic map, quickly became widely influential and formed the basis for numerous calculations of forest cover and deforestation in the protectorate. Citing Emberger's 1938 map, an influential botanist wrote in 1941, for example, that "Morocco had, during the Roman epoch, possibly 65 to 70 percent of its surface covered in forests."[110]

Emberger's research—and especially his exaggerated estimates of naturally (potentially) forested regions of Morocco, and thus deforestation rates—had a profound influence on the director of forestry, Paul Boudy. Emberger's estimates gave Boudy the authority of science to legitimize his preexisting views on the forests in Morocco and North Africa. This becomes clear in the first volume of his magisterial four-volume series, *Économie forestière nord-Africaine.* Although published in 1948, important parts of volume one, on the human and physical milieu, were completed by 1941, only two years after Emberger's map of bioclimatic regions appeared. Boudy devoted significant space in this first volume to introducing and discussing Emberger's stages. He lauded their precision, writing that "Emberger's method provides . . . a precise scientific basis for classifying diverse types of forests."[111] Boudy then used the Emberger method and map to calculate the amount of forest that had disappeared since classical times "by pure ecology."[112] In this section, Boudy calculated that about six million hectares should naturally/ecologically be covered with forests, although only about 4.3 million hectares actually were forested. Thus he concluded that Morocco was between 53 and 58 percent deforested.[113] Based on these calculations, approximately 25 to 30 percent of Moroccan territory should be forested, while only about 15 percent was actually forested. Boudy did not, however, limit himself to Moroccan forests. He also applied the Emberger method to Algeria and Tunisia.[114] At Boudy's request, Emberger created a map of the stages of vegetation for Algeria and Tunisia to make a complete set of bioclimatic maps for French North Africa.[115] For Algeria he then used Emberger's new bioclimatic map and Maire's phytogeographic map of Algeria and Tunisia to calculate, using "the climatic surfaces and principal associations," that Algeria had lost about five million hectares of forest and thus was approximately 63 percent deforested.[116] He used the same maps and methods of calculation to determine that Tunisia was about 56 percent deforested. Overall, he concluded that the Maghreb had lost at least ten million hectares of forest over the previous centuries and was more than 50 percent deforested.

A dozen pages later, Boudy provided a detailed history and description of how this deforestation and desertification had proceeded over the last two thousand years, since the time of "Roman Africa." This section reads like a textbook presentation of the declensionist environmental narrative of North Africa as conceived under French administration—complete with facts, figures, and maps. Boudy began by suggesting that, according to ecological science, all of the areas that should be forested naturally were actually forested during Roman times. Although he conceded some forest destruction in North Africa from about the fifth century, he placed the greatest blame on the eleventh-century Arab invaders and their destructive herds, often citing Ibn Khaldoun. This "army of locusts," "this plundering and anarchic horde," "had the effect of intensifying the practice of nomadism . . . and of transforming the country into desert."[117] With Emberger's new ecological research, Boudy was able to justify the old narrative with "scientific facts." In this way, Boudy was a pivotal actor in refining, institutionalizing, and amplifying the declensionist narrative in Morocco and, more widely, in Algeria and Tunisia.

Restoring the Range: Narrative, Policy, Practice

Both Boudy and Emberger were therefore strongly influenced by the declensionist environmental narrative, and, in turn, their quasi-scientific stories of deforestation reinforced and rationalized the French colonial vision of the North African environment as being deforested, overgrazed, and desertified. What had been largely a literary story of generalized environmental decline for more than a century became, by about 1940, a scientific story complete with ecological statistics and maps that helped to justify policy formulation. In Morocco, the influence of this narrative reached beyond the realm of forestry, however, to the policies and work of the Livestock and Range Service (Service de l'Élevage), which governed most of the unforested areas of the protectorate. Notably, the Forest Service itself played a large role in the environmental policies developed in the Livestock and Range Service. In Algeria and Tunisia the pastoralists, by the early twentieth century, had been mostly sedentarized and their migration routes had been rationalized and regulated. Permits like those shown in figure 5.8 were required for pastoralists to travel in most parts of Algeria and Tunisia. In Morocco, however, a large number of pastoralists migrated widely in the mountains, deserts, and steppes, with little regulation, early in the protectorate period. Boudy, among others, considered the issue of pastoral migration of widespread importance in Morocco, but thought that in Algeria and Tunisia migration had only a limited range and thus was less important. Range management was developed in Morocco at about the same time that the declensionist narrative was being refined and quantified by Emberger and Boudy, which makes the case of Morocco unique in French North Africa. Although many of the policies and techniques developed and applied in Morocco were

Figure 5.8. Migration permits for nomads, Tunisia. These permits required by the French government in Tunisia for the migration of nomads and other pastoralists are similar to those required in Algeria. Names, "tribal affiliations," numbers of humans and animals, origin and destination, weapons, and medical observations were required information. Permission to travel was granted only for specific routes and destinations. From Centre des Archives Diplomatiques de Nantes, France, "Tunisie," Premier versement, 876, Transhumance. Reproduced with the kind permission of the Archives du Ministère des Affaires Étrangères, Nantes, France.

similar to those in Algeria and Tunisia, the Moroccan experience provides a concise case study of how the colonial narrative strongly affected most of the territory outside of forested and agricultural sections of the Maghreb.

Boudy's early conviction as to the terrible damage inflicted by livestock grazing in the forest had a strong impact on forest policy, but it also influenced livestock and range management. This negative view of grazing was widespread among protectorate administrative officials and colonists alike.[118] Grazing in the forest, and especially the practice of transhumance/migration, a centuries-old pastoral livelihood tool, were seen as the greatest causes of deforestation in Morocco and therefore as the greatest threats. Under Boudy's direction, the Forest Code was amended in 1921, with a new provision regulating the right and usage of pasture within forested areas. The amendment limited grazing rights to the "traditional users" of the forests, thereby effectively regulating migration.[119] Traditional users were defined as members of "Moroccan tribes" who lived in the territories where the forests were located or who had had use rights to the forest for a very long time. This definition of traditional users eliminated large numbers of local people and animals who had actually been using the forests because of the way the French defined, delimited, and "fossilized" the "Moroccan tribes."[120]

The numbers of livestock were strictly limited by defining the size of the family herd allowed to graze tax-free in the forest. Although relief from taxes was a benefit, the family herd could include, according to the new law, no more than five cows and fifteen sheep, and no goats or camels.[121] Typical herds in much of Morocco, especially of sheep and goats, were much larger than this. The family's other animals could graze also, if the annual limits were not exceeded, but they had to pay a fee per head of livestock to be admitted. These annual limits, in turn, depended on the Forest Service's annual assessment of the state of the vegetation in each forested area (*le possibilité en herbe*). Grazing was still limited to areas judged defensible, and the routes through which the livestock could travel to get to the permitted sections of forest were strictly delimited and monitored. Furthermore, users had to reapply each year for the right to graze their herd in their traditional forest. Livestock grazed in the forests were also subject to the *tertib*, the agricultural-production tax.[122] Overall, this resulted in a large reduction in numbers of livestock grazed and diminished rural Moroccans' abilities to support basic family needs.

Whereas fire and incendiarism had been significant problems facing the forestry administration in Algeria, grazing was considered a more important problem in Morocco. According to Boudy, grazing was "the nerve center of the forest question in North Africa."[123] His solution to the problem was to combine repression with technical organization. As he put it, "surveillance, and consequently repression, must continue to be exercised in the most active fashion."[124] During the 1930s the government paid heightened attention to the issue of migration among pastoralists in Morocco and its assumed contribution to overgrazing. In 1934, new regulations officially reduced the territories and rights of usage of mobile pastoralists as well as the numbers of stock that could migrate in several regions. This not only followed the final pacification of the protectorate—specifically, the conquest of the south, with its high proportion of nomads—but also increased concern about the migrations of livestock in Algeria and in France. This decade saw the publication of many articles on the subject, almost all of which argued for the diminution if not the elimination of migration in Algeria, France, and Morocco, largely for environmental reasons.[125] Where it was thought that migration could not be eliminated quickly, strict surveillance and control over all movements were considered essential.[126] In 1939, the government officially acknowledged "that the goal to attain is the progressive extinction of transhumance [migration]."[127] Although overgrazing and deforestation were given as the primary reasons that migration was a problem and needed to be stopped, the protectorate administration was also interested in controlling or eliminating migration for reasons of security and to quell the rising tide of nationalism in the protectorate. The Middle Atlas region, in particular, was believed to harbor several groups thought to be sympathetic to revolution and independence from France.[128]

Forestry policies thus overlapped with and influenced policies governing livestock raising and range management in Morocco. The Livestock and Range Service, a division of the Agriculture Department, was founded in 1913, the same year as the Forest Service.[129] Until 1930, it was directed by veterinary colonel Théophile Monod (1864–1942), a good friend of Lyautey's. Monod was born and educated in France, had many years of experience in Algeria and elsewhere in greater France, and stayed active in an advisory capacity for twelve years after his retirement, until his death in Casablanca.[130] In the early years the service was concerned primarily with the control of infectious livestock diseases and the immense task of trying to transform subsistence production, especially of sheep and cattle, into commodity production. Toward that end, the service worked to intensify and "improve" livestock raising to produce more meat and milk, as well as wool of a higher quality, that would appeal to the French consumer. This involved improving nutrition, providing shelter from heat and cold, and reducing physical exertion, especially reducing travel over long distances. This mode of livestock raising was completely foreign to the majority of Moroccan subsistence producers; it was also much more expensive.

This sort of intensive livestock production, which could be accomplished in a smaller area with less movement, was also thought to be necessary as the protectorate developed its agriculture and forest sectors. Just as large areas of the forests were increasingly being protected from grazing, more and more land was being brought under cultivation, much of which had been previously used as seasonal pasture land. Thus the Livestock Service worked in cooperation with the Forest Service and Agriculture Department. It also had to adapt to the changes wrought by forest policies that progressively decreased pasture land and migration in the forested areas of the country. In 1928, for example, grazing was restricted on two million hectares of prime pasture land when these areas—dominated by alfa grass (*Stipa tenacissima*)—were placed under the control of the Forest Service and new legislation regulating their use was promulgated.[131] In the early 1920s the Livestock Service also became responsible for pasture (range) improvement. As Monod explained, the Livestock Service was charged with the amelioration of the whole environment in which animals were raised, and therefore "also with obtaining the reconstitution of pastures."[132] The service organized, run, and staffed by veterinarians therefore was responsible for what today is known as range management. A vast majority of the land in the protectorate not under cultivation, not in forested areas, and not in urban or village areas was used for grazing livestock and thus was frequently subjected to the policies of the Livestock and Range Service.

Within a decade of the establishment of the protectorate, the Livestock and Range Service was implementing many policies to reconstitute pastures assumed to be degraded. The conventional environmental narrative is evident in this sector, too, with numerous references to the natural fertility of Moroccan pastures that had been degraded by the "improvidence of the natives," especially transhumants and nomads.[133] Although the degree of pasture degradation never was as tightly quantified as that of deforestation in the 1930s and 1940s, it was the target of dire claims made by the department. By the 1940s, the word *désertification* was regularly used to describe what French officers working in southern Morocco believed to be the severe degradation of pastures. In 1949, Captain Salvy reported that in southern Morocco the nomads had overgrazed and exhausted the pastures and thus set in motion "all the processes of desertification."[134] Two years previously, Captain Azam had detailed these processes of desertification when he claimed that the nomads in the Draa region of southern Morocco had, for centuries, caused "the end of the forest, surrendered without hindrance to the flocks to the point that, according to the profound phrase of Ibn Khaldoun, 'the soil itself seemed to have changed in nature.'"[135] Based on the recommendations of reports like these, techniques to reconstitute and restore the range were implemented widely, including delimiting exclosures for forage reserves, seeding pastures with forage species, creating cactus plantations in hyperarid regions and irrigated alfalfa pastures in those areas with access to irrigation networks, and acclimatizing exotic forage species.[136] Later in the protectorate, foreign (presumed better) forage shrubs such as Australian saltbush (*Atriplex*) were planted to improve pastures. Pastures improved by

Figure 5.9. The High Plateaus, Algeria. The caption reads, "As far as the eye can see, the steppe, flat, bare. A few scarce tufts of grass nourish the herds of the nomadic tribes. Landscape of desolation, that already announces the Sahara." Images with captions such as this were common and helped to strengthen arguments that areas like these needed to have their pastures "improved"—and that the activities that had apparently led to the denuding of the vegetation had to be curtailed. From Jean Blottière, *Les Productions algériennes,* Cahiers du centenaire de l'Algérie (Orléans: Imprimerie A. Pigelet et Cie., circa 1931), 11.

such means, though, often required paid permits before entry was allowed, and grazing was restricted and regulated. Many attempts at improvement failed when imported plants did not grow well.

Most of these measures discouraged mobility. They required time, money, and/or work spent in a single location; that is, they encouraged and often enforced sedentarization. These so-called improvements were also largely in complete opposition to the primary goals of subsistence herders. Poor herders need high mobility and hardy livestock. Hardy livestock, though often low producers, allow greater flexibility of resource use in an unpredictable environment.[137] The protectorate's overriding goal of security helped to reinforce the transformation from a nomadic/transhumant mode of production to a sedentary mode. It is clear that, to this end, the Livestock and Range Service worked more closely with the military and intelligence services than did the Forest Service.[138] Over the course of the protectorate period, the policies of the forest and livestock services greatly curtailed—sometimes brutally—traditional modes of livestock production, both nomadic and transhumant. As a result, an untold number of local Moroccans lost their livelihoods, and many migrated to the shantytowns of the growing cities.

The delimiting, improving, and policing of pasture lands was not as comprehensive or as detailed as similar activities were in forested areas. The policies and work of the Livestock and Range Service, however, did help to extend many of the goals developed by the Forest Service—and the environmental narrative—to much of the rest of the protectorate. As it had in the forestry sector, the narrative served colonial interests in the livestock sector much more than indigenous interests. In this sector, a combination of policies (encouraging intensification, veterinary public-health laws, and range-management practices that enforced sedentarization) privileged commodity production by European colonists over the subsistence production of local Moroccans. This had the added benefit, in the eyes of the military and parts of the administration, of controlling the nomads and other mobile herders who were perceived as a threat to the state.

Those in need of labor also benefited. By incrementally depriving Moroccan pastoralists and agropastoralists of the ability to support themselves with forest products and livestock raising, the government's policies created a vast pool of labor for the expanding export-oriented agricultural venture that defined Moroccan colonialism. Such labor was necessary, since by the end of the protectorate period an estimated 77 percent of the best agricultural land was in the hands of Europeans, mostly in large farms requiring many workers. The deforestation/desertification story was invoked here, too, with an ironic twist. As Boudy explained in 1927 to the young recruits of the government's Indigenous Affairs Service, the argan forest (in southwestern Morocco) had to be saved, for "if the argan [tree] disappears, the region will return to its true steppe destination or even to desert; with it will be dried up the great human reservoir of Morocco, which feeds the rest of the country with manpower."[139]

This kind of transformation had already occurred in Tunisia and Algeria, where, decades earlier, large segments of the rural population, especially pastoralists, had been dislocated and forced into wage labor by colonial policies. The colonial environmental narrative, though, newly quantified and institutionalized in the form of potential vegetation maps, deforestation statistics, and estimates of desertification, did find new applications in Tunisia and Algeria as well as in Morocco. In the 1940s and 1950s, for example, a new service, the Service for the Defense and Restoration of Soils (Défense et Restauration des Sols [DRS]), was created in Algeria (1941), Tunisia (1946), and Morocco (1951).[140] The DRS was housed within the Forest Service in all the French North African territories. Its purpose was to ameliorate or prevent soil erosion presumably occurring as a result of deforestation and desertification. The statistics on deforestation upon which these offices acted were nearly entirely drawn from the conclusions of Boudy and of others influenced by his work. Using these statistics, and invoking the colonial declensionist environmental narrative, the DRS made technical "improvements," such as banquettes, reforestation perimeters, eviction of local peoples, interdiction of grazing, and prohibitions on most uses of many types of vegetation on about one million hectares of land over the next thirty to forty years. Not only did these programs encounter resistance from most North Africans, especially the poor, but they also failed to curtail erosion and actually exacerbated it in many places.[141]

In the 1980s, this kind of work, informed by the same colonial narrative and statistics, was mostly taken over by programs designed to halt desertification, many of them financed and planned by international institutions and NGOs. Thus, the institutionalized declensionist narrative influenced environmental policies across the Maghreb up to and beyond independence, continuing into the twenty-first century.[142]

CHAPTER 6

Decolonization, the Colonial Narrative, and Environmental Policy Today

THE SPURIOUS COLONIAL STORY of North Africa's long environmental decline at the hands of the "natives" informed and motivated much of the French venture in the Maghreb for over a hundred years. This declensionist narrative incorrectly blamed the North Africans, and especially pastoralists, for deforesting and desertifying the former granary of Rome, following the "Arab invasion" of the eleventh century. Building on an earlier European narrative highlighting the fertility of the Maghreb, the environmental narrative was variously shaped, early in the colonial period, by different actors for a wide variety of reasons. It was used to justify property laws that disenfranchised the North Africans and facilitated the sedentarization of nomads, to instigate reforestation schemes, and to justify draconian forestry laws, in addition to being actually incorporated into these laws. Late in the colonial period the narrative was rationalized, formalized, and institutionalized in the young science of plant ecology, a key development that ensured the long legacy of the narrative. Throughout the colonial period, this declensionist narrative served three primary purposes:

the appropriation of land and resources; social control (including the provision of labor); and the transformation of subsistence production into commodity production. In all of these areas, French colonial interests almost invariably triumphed over indigenous interests.

In Algeria, for example, near the end of the colonial period (in the early 1950s), the European population accounted for about 10 percent (984,000) of the total population, and only 3.5 percent of the agricultural population, yet it controlled approximately 38 percent of the best agricultural land (2,818,000 hectares), with an average of ninety hectares of land per person.[1] The Algerians, accounting for about 90 percent (8,546,000) of the total population, owned 62 percent of the arable land, with an average of only thirteen hectares of land per person.[2] Similarly, 75 percent of irrigated crop land (the most productive agricultural land) was owned by Europeans.[3] Nearly all of this arable land was located in the northern part of Algeria. In addition to the nearly three million hectares of land owned outright by Europeans (including 211,000 hectares of forest), the Algerian state and the communes owned a further 7,235,000 hectares, constituting about half of all the land in northern Algeria.[4] The Algerians owned only 12.5 percent of forested lands, the state owned 72.5 percent, and the final 15 percent were owned by colonists and large European companies.[5] In the pastoral sector, the nomads had been reduced to only 5 percent of the population, whereas they had accounted for about 60 to 65 percent of the population in 1830.[6] Livestock owned and raised by Algerians declined significantly between the 1880s and the 1950s. Sheep owned by Algerians, for instance, were estimated at about 10.5 million in 1887 but only at about 3.8 million in 1955.[7] An unknown amount of the best grazing lands had been appropriated for colonial agriculture.

At independence, the situation was similar in Tunisia, where the Europeans, only 6.7 percent of the total population of 3.8 million, owned 22 percent (850,000 hectares) of the best arable land in the country.[8] The average French holding was 250 hectares while the average Tunisian holding was only six hectares.[9] The Tunisian government owned all of the 754,000 hectares of forested land, since communal forests for the Tunisians were not recognized by the state.[10] Nearly all the nomads in Tunisia had been forced to sedentarize. One French scholar has described this inexorable process as one in which the pastoralist way of life "was degraded into a miserable sedentarization."[11]

In Morocco, just prior to independence, the European settlers represented about 4 percent of the total population (8.6 million) but owned about 15 percent (1,017,000 hectares) of the best agricultural land.[12] The average European holding was about 170 hectares, whereas the average holding for most Moroccans was less than five hectares.[13] Although Europeans accounted for only 0.5 percent of Morocco's landowners, they owned 47 percent of all irrigated land (35,800 hectares) in the protectorate.[14] As in Tunisia, all forested land was claimed by the state, and thus 4.8 million hectares of land classified as forest was reserved and regulated by the Forest Service, which allowed only limited access by a few traditional users.[15] The nomads and seminomads of Morocco suffered from the campaigns of pacification, during which many animals were killed, and from the expansion of colonial agriculture into the best grazing lands. As a result, the seminomads of the plains and mountains were largely sedentarized.[16] Although the nomads of the eastern High Plateaus and the southern territories retained much of their way of life at independence, their traditional migration routes, especially to the north and west, were denied to them. Despite these disruptions, Morocco retained the highest proportion of nomads of any of the Maghreb countries, and an estimated 17 percent of the rural population was nomadic as late as the mid-1970s.[17]

The disenfranchisement of the North Africans from their lands and resources resulted in the enrichment of many European settlers and businesses (and of a small minority of elite North Africans) and the mass impoverishment of nearly twenty million indigenous people. Levels of poverty and unemployment were particularly acute immediately prior to independence in the mid-1950s.[18] In Morocco, about 42 percent of Moroccans were landless—most of them in rural areas, but with significant numbers also concentrated in shantytowns (bidonvilles) in the large cities. Estimates for this period are that about half of the Moroccan agricultural population was significantly underemployed. The average per capita annual income for a Moroccan was 31,000 francs, while it was 590,000 francs for a European, twenty times greater.[19] Land concentration and poverty were serious problems in all three colonies as independence approached. In Tunisia, the average per capita annual income for Tunisians was about 16,000 francs, and nearly one-third did not have permanent employment.[20] The Europeans in Tunisia are estimated to have had an annual income at least six to seven times greater than the average Tunisian.[21] In Algeria, average per capita annual income for the indigenous population was approximately 24,777 francs, while for Europeans it was 414,000 francs, about seventeen times greater.[22] It is estimated that up to half of the rural Algerian population of working age was unemployed in the mid-1950s.[23]

Deprived of land and resources and swept into money economies by taxation, many North Africans were forced into wage labor on European agricultural estates, or migrated to France in search of work there. The resulting disruptions to traditional society and social structures are inestimable.

Colonialism's effects on the physical environment of the Maghreb were equally profound. As colonial agriculture spread and became mechanized, more and more marginal land (land receiving less than three to four hundred millimeters of annual rainfall) was plowed and planted, much of it for grain production. Most of this marginal land had previously been seasonal pasture, and it did not reliably produce crops without irrigation. When grains were planted, productive harvests were obtained only intermittently, since the precipitation in North Africa is very irregular and unpredictable. Disturbing these marginal soils with heavy, European-style plows actually desiccates them, since breaking the surface and turning over the soil increases evaporation of soil moisture and significantly reduces crop yields over time. By contrast, the traditional scratch plow used by the North Africans before the colonial period did not cut as deeply into the soil and was used at different times in the agricultural cycle, and thus did not expose as much of the underlying soil moisture to the heat and aridity of the Maghreb environment. The traditional scratch plow also helps to prevent soil erosion under these conditions, whereas the European plow does not. Additionally, land concentration forced a reduction in fallow periods, which in arid environments are as crucial for restoring soil moisture as for restoring nutrients. Overall, the introduction of "modern" European types of agricultural techniques, the reduction of fallow periods, and the spread of agriculture onto marginal soils produced perfect conditions for soil degradation (desiccation and erosion) in a large number of areas during the colonial period.[24]

To make matters worse, perennial irrigation and the proliferation of large dams during the colonial period led to problems with overirrigation, salinization, and the depletion of groundwater, all of which would be greatly exacerbated in the postcolonial period. In the High Plateaus and steppes, the overextraction of alfa grass for paper production led to the degradation of large areas of formerly productive grazing land, especially in Algeria.[25] The forests of the Maghreb also suffered under colonial administration. An unknown number of hectares of forests were destroyed during the wars of occupation, especially in Algeria and Morocco. The occupying armies burned and cut down trees to try to control the resisting populations, while consuming wood for their own daily needs. The harvesting of cork, timber, and other forest products for export also took their toll.

Although it is impossible to calculate with any accuracy the losses to North African forests as a result of French colonialism, there have been some estimates. Forester Paul Boudy, for example, estimated that in Algeria alone "at least 1,000,000 hectares of the wooded surface disappeared from 1870 to 1940."[26] Much wood was also procured from Maghreb forests by the French during World War I and World War II. The wars of independence, particularly in Algeria, were also very damaging to the environment. In Algeria, "hundreds of thousands of hectares of forests were burned or defoliated by napalm bombs; cultivated lands were either sown with mines or declared 'prohibited zones;' the country's livestock was almost decimated."[27] The results of increased urbanization and industrialization during the colonial period, especially in the coastal zones of the Maghreb, further modified the Maghreb environment, not least by contributing toxins and pollutants to the soil and the limited water supplies.

Despite the growing evidence, even during the colonial period, that much of the declensionist environmental narrative was weakly supported and that large transformations of the environment had taken place as a direct result of colonial actions, the narrative detailed in this book remained ubiquitous. By the end of the colonial period it had become the dominant environmental history of the Maghreb. It had been embedded in numerous important and influential publications, including the forestry manuals, histories, botanical and agricultural treatises, and geographical and ecological studies. Perhaps not surprisingly, the French military invoked and perpetuated the narrative in some of its own publications for military personnel published during the Algerian war of independence.[28] Many of these authoritative colonial works continue to be read widely during the postcolonial period and thus have laid the foundation for a great deal of subsequent education, research, and policy formation.[29] Rather than questioning or even examining the colonial declensionist environmental narrative, the postcolonial states have embraced it. Indeed, this narrative has proven to be very useful for the newly independent governments to support various development projects, and for dealing with segments of their population that they deem "difficult." In both Morocco and Algeria, for example, the declensionist narrative has been invoked by postcolonial governments to justify projects that have disenfranchised pastoralists in the name of stopping "desertification."[30] One of the most notorious is the *barrage vert*, or "green dam," in southern Algeria. This project "to hold back the spread of the Sahara" by planting 1,500 kilometers of desert with trees began in the 1970s. The pastoral nomads who lived and migrated in the region were forced to settle outside of the area.[31] The green dam is now considered an ecological failure and the trees that were planted have had a very low rate of survival.[32] Morocco, meanwhile, is planning a green dam south of the city of Dakhla in the Western Sahara, with similar ideological motivations.

The colonial narrative has also been perpetuated, in whole and in part, in scholarly and popular writing about the Maghreb in France and in much of the world to the present. Most printed sources that treat the environmental history, or changes in the environment over time, in some way cite these colonial sources, especially the work of Boudy, Emberger, and Bernard. It has been incorporated in histories of the region as well as in environmental and scientific works.[33] A particularly widely cited example in English is the Canadian forester J. V. Thirgood's book, *Man and the Mediterranean Forest.* Thirgood claimed that "today, North Africa, the fertile granary of the Roman Empire, is largely desert or semidesert."[34] He explained immediately prior to this that the destruction was wrought primarily by the Arab nomad "invaders." Later he cited Boudy's deforestation statistics to highlight the amount of deforestation that has occurred, compared to the potential forest that could grow in the absence of pastoral people and their herds. Thirgood also cited the work of the botanist/ecologist Emberger. Colonial sources continue to be regularly cited for their desiccationist and other spurious ecological arguments as well. One recent article by an Algerian cited Boudy to support his claim that "in North Africa . . . forests have a decisive influence on the ecosystems and the local climate; increasing rainfall by an average of 8 percent."[35] This author opened his article with a quote from Thirgood relating that North Africa "used to be described as 'a land of continuous shade,'" and used the deforestation figures Thirgood had taken from Boudy.[36] Other examples abound of the incorporation of the colonial narrative, especially from the environmental literature, where the narrative still appears to be especially strong today.[37]

Indeed, the work of colonial scientists like Emberger, Maire, and Boudy, their authoritative calculations of potential vegetation, and the resulting exaggerated estimates of deforestation and land degradation in North Africa, have been among the most influential and long-lasting legacies of the declensionist narrative. Their descriptions and maps of potential vegetation and their calculations of rates of deforestation (and its causes) are invoked frequently as scientific proof of the state of the North African environment, and thus provide legitimacy for many different projects. Despite the fact that the paleoecological and contemporary ecological evidence does not support these exaggerated estimates of deforestation and land degradation in North Africa, postcolonial governments and international institutions such as the United Nations continue to use such colonial statistics to justify and plan many development projects in the region. These colonial statistics have also been used to generate other degradation indices, like desertification estimates, that have frequently been shown to be exaggerated.[38] The idea of desertification itself is in fact a colonial construction, a concept with little basis in empirical evidence initiated and propagated by those with a poor understanding of arid-land ecosystems.

This book has shown that the origins of both the word and the idea of desertification have deep roots in the French colonial Maghreb, reaching back to the early nineteenth century. The word and concept have nevertheless been most frequently traced to the 1949 work of André Aubréville on tropical forests.[39] The word seems, however, to have been first used in French in 1927, when the forester Louis Lavauden wrote that the forests of the Sahara were "desertified" (*désertifiée*) and stated that desertification "is uniquely the act of humans. . . . The nomad has created what we call the pseudo-desert zone."[40] The concept of desertification, though, appears as early as the 1830s in Algeria, when blame was placed on nomads for ruining land in order to excuse taking their property.

As the colonial period progressed the idea of desertification became widespread and discussions of it took on an increasingly urgent tone. In 1880 a prominent Algerian forester argued for massive reforestation, for instance, by claiming that the nomads "had created the dunes and the sand" in southern Algeria.[41] Crisis narratives of an enlarging Sahara creeping north were increasingly common from midcentury. Paulin Trolard, the influential director of the Ligue du Reboisement de l'Algérie, warned that the Sahara would soon engulf and destroy all of Algeria if action was not taken. The "crisis" of desertification lay behind the numerous projects already discussed that worked to the disadvantage of nomads and other pastoralists. It had a tenacious staying power and gained strength in the early twentieth century. The celebration in 1930, for example, of one hundred years of French rule in the Maghreb provided the occasion for the publication of many books in which descriptions of desertification were prominent. The author of one of these, *La Tunisie,* invoked the narrative when he described the south of Tunisia as "desertified" (déser-tifiée) as a result of human action.[42] In 1949 Captain Salvy used the word desertification in his report on the crisis of nomadism in southern Mo-rocco and cautioned that "the Sahara is advancing towards the North and towards the South."[43] The postcolonial governments in North Africa have continued to invoke crises of desertification, for a variety of reasons.[44]

The United Nations also helps to perpetuate the colonial crisis narra-tive of desertification and deforestation in many of its publications. One recent article, in its discussion of the imminent dangers of desertification, stated that "the Roman Empire's bread-basket in North Africa, which once contained 600 cities, is now a desert."[45] The legacy of the declen-sionist narrative is also, as could well be expected, very strong in the "anti-desertification industry" in North Africa today.[46] Thus the narrative per-sists, despite convincing evidence that most of the Maghreb has not been significantly deforested or desertified by burning or overgrazing, and has in fact never been heavily forested during the last three thousand years. Overgrazing is the single factor most commonly blamed for desertifica-tion in the region, and it is most often the livestock of pastoralists that are singled out for reform. This is so despite the fact that the local ecol-ogy is very well adapted to fire and grazing, both of which were used sus-tainably in the precolonial period and could be again. Nonetheless, mil-lions of dollars of international aid money and sparse national resources are spent on fighting desertification in the Maghreb with programs that emphasize lowering livestock numbers and regulating the movements of nomads and their herds. The documents associated with these projects fre-quently invoke all or part of the colonial declensionist environmental nar-rative, including calculations of current or past deforestation rates that rely on colonial estimates. Very few scholars have investigated these claims of degradation or sought to evaluate the many bodies of evidence available, and none have explored the colonial precedents of the degradation claims.[47]

Scholars of the Middle East and North Africa have, in fact, been no-ticeably silent on these topics. These lacunae have been noted by recent scholarship in environmental history.[48] Although many important works have appeared over the last decade analyzing the environmental history of sub-Saharan Africa and South Asia, research on the environmental his-tory of the Middle East has been very sparse: only three major monographs have been published over the last thirty years.[49] None of them treat the Maghreb. This book is the first to explore the environmental history of North Africa and particularly its complex relationship with colonialism.

Thanks to recent work in critical environmental history and political ecology in sub-Saharan Africa and Asia, false environmental narratives, repressive environmental policies, political and economic struggles over resources, and interest-laden narratives are now well recognized phenomena.[50] Although nearly all of the research on Africa traces dominant environmental narratives back to the beginning of the colonial era, few explore the nineteenth century in any depth, since most colonial ventures in Africa began only late in the nineteenth century or later. Nor do many of these works recognize that the precedents for French sub-Saharan territories were set by the experiences of the French in North Africa, particularly in Algeria.

The Maghreb territories, though, especially Algeria, were in fact held up as "models of colonial installation" to be emulated in other French territories in Africa.[51] The research presented here suggests that the influence of Maghreb environmental narratives and environmental policies in French colonial Africa as a whole has been overlooked and should be explored. To date, most intellectual histories of environmental thinking have traced ideas from French, British, and European precedents to the Indian subcontinent and subsequently to sub-Saharan Africa. Yet the evidence shows that many common colonial arguments regarding deforestation, desiccation, desertification, and the "improvidence of the natives" have deep roots in early nineteenth-century Algeria. Moreover, several innovations in colonial policy and administration likely came from French experiences there. For example, several of the eighty-one British foresters trained at Nancy between 1867 and 1893 went to work in India, and probably transmitted key concepts such as the taux de boisement or the need for at least one-third of a country to be forested to be inhabitable by "civilized" people. It is worth noting that French publications on forestry that included information on Algeria were translated into English and published in India for British foresters working there.[52]

The importance of the French declensionist environmental narrative of the Maghreb for colonialism and for environmental history has also been overlooked by French scholars. In fact, French scholarship is largely lacking in the area of environmental history, with a few exceptions.[53] Neither historians nor geographers appear to have produced much research on any of the major themes that are evident in the work of scholars of environmental history from other parts of the world. Yet French colonial environmental narratives played a crucial role in the colonial project in North Africa. The declensionist narrative greatly facilitated the dispossession of large numbers of North Africans from their land, and the criminalization of many of their traditional activities in the forests and the deserts, in ways not previously recognized even by postcolonial scholars deeply committed to issues of social justice and historical accuracy.[54] The multiple ways, however, in which the declensionist narrative was used and the methods by which it was incorporated into colonial legislation on property, forestry, and range management have exerted an enduring influence on environmental policies in the Maghreb, several of which survive to the present. The colonial 1917 Forest Code in Morocco, for example, remains the forest code today, with a few amendments that have been added over the years.

The potential vegetation maps created by French colonial botanists and ecologists in the Maghreb, and their use by foresters and others to generate environmental data sets like deforestation statistics and desertification estimates, may represent the most important legacy of the declensionist narrative. Boudy's work continues to be cited regularly, nearly sixty years after publication. At least twenty-seven academic papers have cited his *Économie forestière* in the last fifteen years alone (six of these in the last two years). His statistics are frequently cited in United Nations project documents as well as in the documents of other international environmental institutions and NGOs.[55] The work of Emberger has also been very influential and long-lived in this area.[56] Emberger's research included work on the climate and vegetation of the Mediterranean basin, and he produced with Henri Gaussen the widely used 1962 *Bioclimatic Map of the Mediterranean Basin* for UNESCO/FAO.[57] This map and others of Emberger's publications also formed part of the basis for UNESCO's subsequent *Vegetation Map of the Mediterranean Zone*.[58] In turn, both of these maps provided much of the information on the Mediterranean zones for UNESCO's *Vegetation Map of Africa*, which remains the most authoritative vegetation map of the African continent today.[59] It is little known, however, that Emberger defined the Mediterranean climate zones and therefore its vegetation zones on the example of Morocco, which he believed alone had all the forms of the Mediterranean climate.[60] The declensionist narrative that so permeated his research on Morocco and resulted in greatly exaggerated estimates of potential vegetation and deforestation rates was then generalized, to some degree, to the Mediterranean basin in these influential maps of climate and vegetation.[61]

Although a few scholars have mentioned the importance of potential vegetation maps in their discussions of colonial forestry, none have yet provided detailed examinations of how these maps were constructed or by whom.[62] Few of these maps have been evaluated for their accuracy in light of contemporary ecological or paleoecological data. None, to my knowledge, have been analyzed for the presence of embedded colonial environmental narratives, despite the fact that these maps provided authority for draconian environmental policies during the colonial period as they still do in much of the world. Many potential vegetation maps used today were constructed during the colonial period or are based on slightly revised versions of colonial maps. These maps and related environmental data sets, such as deforestation estimates, frequently provide baselines for work in remote sensing and geographic information systems (GIS) and thereby influence a large and growing number of environmental projects, some global in scope.[63] Numerous such environmental projects disenfranchise local peoples and criminalize their subsistence activities, or relocate them altogether, in much the same way that similar colonial ventures did decades ago.

Interrogating the deep origins of ideas and constructions like desertification, potential vegetation maps, and their related data sets suggests that we are, in the recent words of one environmental historian, "past the point where we can blithely believe we are generating a positivist picture of empirical reality."[64] Rather, we need to investigate such concepts, maps, and other artifacts of the colonial period, and the faulty narratives that often inform them, to be able to write environmental histories that fully account for the power relations created when the maps and data sets were made. Failing to do so will likely produce more projects, like the "green dam" in Algeria, that foster greater social and ecological harm than good.[65] By doing so, however, by telling stories about stories about nature over the *longue durée,* we may find that we are able to expose and ameliorate the exploitation of less powerful peoples in the name of environmental protection that has been ongoing since the colonial period.

A Note on the Geography and Ecology of the Maghreb

THE MAGHREB IS UNIQUE among North African regions because of its extended areas of elevation and its northerly latitude.[1] Due in part to these attributes, the Maghreb receives more precipitation than the other countries in North Africa, Libya and Egypt, and consequently enjoys a larger proportion of arable land. Stretching across all three countries like a backbone, from the southwest to the northeast, is a nearly continuous chain of mountains. These are the Atlas Mountains in Morocco and the western two-thirds of Algeria, and the Aurès Mountains from eastern Algeria into Tunisia. The highest mountains are found in Morocco, where the Atlas forms three parallel chains, the Anti-Atlas, the High Atlas, and the Middle Atlas Mountains. In addition, Morocco has the Rif Mountains and Algeria has the Kabylie Mountains in the northern areas bordering the Mediterranean. The tallest peaks in the Maghreb are located in Morocco, in the High Atlas, where some are over four thousand meters high. In Algeria, the Atlas Mountains form two parallel chains, the Little Atlas (also called the Tell Atlas) along the Mediterranean and the Saharan Atlas further inland. The Aurès Mountains in eastern Algeria and Tunisia form the highest peaks in these two countries. These mountains and their related highlands cover the greater part of the Maghreb.

The average elevations of Algeria and Morocco are 900 meters and 800 meters, respectively. Tunisia has a lower average elevation of only 300 meters. Both Morocco and Algeria contain significant areas of nearly continuous elevated tablelands running 600 kilometers from eastern Morocco across about two-thirds of Algeria. Elevations range from 900 to 1,300 meters in the western parts of these High Plateaus, but fall toward the east where average elevations reach only around 400 meters. In Algeria the High Plateaus are bounded to the north by the Little Atlas Mountains and to the south by the Saharan Atlas. South of the Saharan Atlas lies the Sahara Desert. In Morocco the High Plateaus are limited to the northeastern section of the country. South of the Atlas are the pre-Saharan desert areas, which have lower elevations. A region of elevated plains called the Steppes is also found in the middle of Tunisia, with the mountainous, humid region to the north and the arid desert region to the south.

These elevated landforms capture orographic precipitation during the winter months when moisture-laden air masses arrive from the north and west and intercept the mountains. Morocco, benefiting from Atlantic storms and air masses, enjoys the highest rainfall in the Maghreb, particularly in the northern Rif Mountains, which frequently receive more than 2,000 millimeters of precipitation each year. Rainfall in most of the country is stochastic, often taking the form of infrequent but intense storms, and amounts vary tremendously from year to year. Summers in the Maghreb are typically hot and dry. Due to the location of the mountains and the prevailing winds during the winter, the most well-watered areas of the Maghreb are along the northern sections, and, in Morocco, in the Atlas Mountains. Most of the major rivers flow from these mountains toward the coast: the Chelif River in Algeria and the Oum er-Rbia in Morocco, for example. Morocco enjoys the most extensive river system in the Maghreb, comprising eight major perennial rivers and several minor streams. Tunisia, by contrast, has only one major perennial river, the Mejerda. In Algeria and Morocco, this combination of factors places the best areas for agriculture along the coastal plains.

The most fertile land for agriculture in Algeria is the Tell, a strip of land across the northern plains of the country, stretching from the coast to the High Plateaus in the south. The "breadbasket" of Algeria—the Mitidja, a former swamp just south of Algiers—is located in this region. The

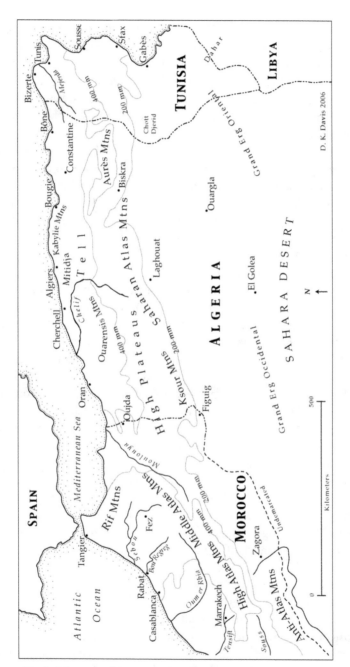

Map 3. Physical features of the Maghreb. Modified after Jean Despois and René Raynal, *Géographie de l'Afrique du Nord-Ouest* (Paris: Payot, 1967), 10–11, 21.

amount of land under cultivation changed over the course of the colonial period, as did the products grown. In 1863, for example, cereals accounted for about 99 percent of agricultural production in Algeria and covered approximately 2,079,612 hectares.[2] Just prior to independence, cereals were grown on 3,276,200 hectares of land, 2,666,860 hectares were under fallow, and various other agricultural products were grown on about one million hectares, bringing the total to around seven million hectares of agricultural land.[3] Much of the expansion in arable land came at the expense of the best pasture land. Several dams built during the colonial period provided irrigation to about 43,119 hectares of very productive farmland by the 1950s.[4] Although it is the largest country in the Maghreb, Algeria has the lowest proportion of arable land. Of its 2,381,741 square kilometers, only about 7.5 million hectares (3.3 percent) are considered arable today, and of these, 569,000 hectares are irrigated. A large area—roughly thirty million hectares—is considered grazing land, and much of the southern part of the country is suitable for occasional grazing depending on rainfall.

The best farmland in Tunisia is found in the valley of the Mejerda River in the north of the country. Tunisia, the smallest of the Maghreb countries, enjoys the largest percentage of arable land. Approximately 30 percent (5 million hectares) of Tunisia's 164,000 square kilometers are currently considered arable. Agricultural land in Tunisia increased during the colonial period, from 800,000 hectares in 1881 to more than four million in 1955.[5] Due to improvements made primarily since the colonial period, nearly 400,000 hectares are irrigated today. Another 2.2 million hectares provide grazing for livestock.

In Morocco, the area of the northwestern plains, from the coast to the foothills of the Atlas Mountains, is the prime agricultural area. The most well-watered country in the Maghreb, Morocco has about 8.5 million hectares of arable land, approximately 18 percent of its 446,550 square kilometers (excluding the Western Sahara). Arable land increased during the colonial period, much of it gained from areas previously used for seasonal grazing. For example, land planted in cereals grew from about 1.9 million hectares in 1918 to 4.3 million hectares by 1955.[6] Although twelve dams were built during the colonial period, only 36,000 hectares were irrigated by independence. Due to a continuation of colonial policies, though, the postcolonial government has enlarged the irrigated area to 1.4 million hectares.[7] About 15.5 million hectares are considered grazing land, but much of the land outside of urban, agricultural, and forested areas is used at some time or other for grazing, as is the case in Algeria and Tunisia.

Without irrigation, agriculture in much of the Maghreb is precarious. Rain-fed cereal harvests vary enormously from year to year, mirroring the irregularities in precipitation as illustrated in figure App.1. Although most of the major rivers in the region have been dammed and many areas subsequently irrigated, the vast majority of agriculture in the Maghreb is still dependent on rainfall.

As a result of the topography and prevailing winds of the Maghreb, those areas south and east of the mountain ranges are located in significant rain shadows, that is, areas of very low rainfall. Approximately 75 percent of the Maghreb is considered arid and receives 350 millimeters or less of precipitation each year. In Morocco, the eastern High Plateaus and the southern pre-Saharan regions receive little precipitation, diminishing from north to south. Whereas the northern Rif Mountains average at least 2,000 millimeters of precipitation annually, the southern pre-Saharan regions are lucky to receive 50–100 millimeters each year. In Algeria precipitation averages 1,000 millimeters in the northeast part of the country, but rainfall diminishes relatively quickly south of the Tell, to 50–100 millimeters and below in the Saharan regions. In Algeria precipitation also follows a general east-west gradient with relatively more precipitation in the

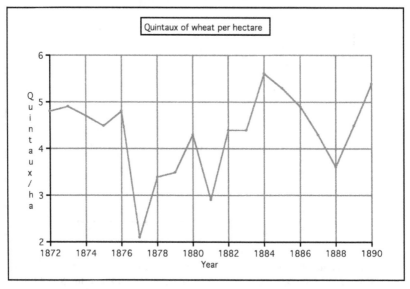

Figure App. 1. Quintaux of wheat per hectare, Algeria, 1872–90. A quintal is equivalent to 100 kilograms. After Charles-Robert Ageron, *Les Algériens musulmans et la France (1871-1919)* (Paris: Presses Universitaires de France, 1968), 1:379 and 2:801.

eastern part of the country and less toward the west. Precipitation is partitioned from north to south in Tunisia, and the northern section of the country is the most well watered. Precipitation throughout the Maghreb is erratic and droughts are common (see figure App.2).

In addition to surface water in the form of rivers and streams, the Maghreb contains significant ground water. All three countries have surface aquifers with varying recharge rates, but only Algeria enjoys large amounts of deep fossil water. These aquifers, located primarily in the Algerian Sahara, were formed thousands of years in the past and have very slow recharge rates. It is these groundwater resources that supply water to the numerous oases in the arid regions of the Maghreb.

The higher rainfall, combined with the moderating influence of the Mediterranean Sea and, in Morocco, the Atlantic Ocean, provides Mediterranean climate conditions that allow Mediterranean vegetation to grow in most of the northern regions of the Maghreb. Much of this vegetation is a Mediterranean-type maquis composed of shrubs and some scrub oaks and other Mediterranean plants such as the dwarf palm. A wide variety of trees also grow, including evergreen and deciduous oaks, cork oaks,

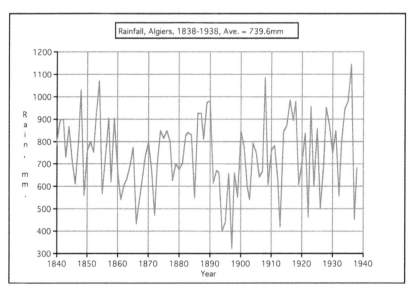

Figure App.2. Rainfall at Algiers, 1838–1938. The 1838–72 data are from CAOM, GGA, P/59, Trottier file; the 1873–93 data are estimated from United Kingdom, Naval Intelligence Division, *Algeria,* Geographical Handbook Series (Oxford: Oxford University Press, 1941), 125; the 1894–1938 data are from B. R. Mitchell, *International Historical Statistics, Africa and Asia* (New York: New York University Press, 1982), 2, 5.

pines, junipers, cedars, and others, many of them found in the mountains. In the high plains of the Maghreb, less rainfall, a longer dry season (six or seven months), higher summer temperatures, and sandier soils result in a steppe vegetation composed largely of artemesias, alfa grass (*Stipa tenacissima*), and other perennial grasses along with a large variety of annual plants. Some of the higher elevations, such as parts of the Saharan Atlas, also have some small clumps of sparse forest composed of trees like pines and junipers. The south of Algeria, more than two-thirds of the country, is desert, as are the southern third of Tunisia and the southern edge of Morocco. In these arid regions that have the least and most irregular rainfall, the highest temperatures, and the highest evapotranspiration rates, desert vegetation predominates. Many annual plants are found in these regions that germinate, mature, and set seed quickly after rare rainfall events. For much of the year, and often for many years in a row, the seeds of these annual plants lie dormant in the soil and will grow only when enough rain has fallen. Perennial vegetation also thrives in the desert. Desert-adapted trees like acacias and shrubs such as tamarisks, oleanders, and jujube trees grow in many areas. Also located in desert regions of the Maghreb are numerous oases where groundwater is used to irrigate a wide variety of crops, most notably the date palm tree. Overall species diversity in the Maghreb is high.

Much of the vegetation in the Maghreb is well adapted to such conditions of the region as aridity, drought, fire, and grazing, and as a result is remarkably resilient to disturbances.[8] The vast majority of plants in the Maghreb, for instance, is adapted to aridity and drought. These adaptations to low rainfall and long periods without moisture are found in both the Mediterranean and the desert vegetation of the region. These plants avoid moisture loss by a variety of mechanisms, including having small leaves that are often covered in minute hairs or waxy coatings (sclerophylls).[9] Other plants can shed their leaves during the dry season, limit their transpiration to the cooler nighttime periods, or absorb dew through their leaves. A great many of these plants have very deep root systems that can tap into moist soil far below the surface. One of the most striking adaptations found in a large group of these plants is the actual avoidance of drought. Annual plants escape drought by having a very short life cycle that begins only after an adequate amount of precipitation has fallen. They germinate, grow, and flower very quickly, sometimes in only a few weeks. After setting seed, annual plants become senescent, but their seeds lie dormant in the soil, sometimes for decades, awaiting the next rainfall to begin the life cycle again. In many parts of the Maghreb over 50 percent of the plant species are annual plants.[10] In these areas, aboveground, visible vegetation is a poor indicator of the health of the ecosystem. In these environments, thousands of seeds are often found per square meter of soil, a rich seed bank that will germinate with the next rain. Even some perennial grasses that normally reproduce vegetatively, such as *Stipa tenacissima* (alfa grass), also have the ability to produce exceptionally high numbers of seeds during periods of heavy rainfall. This research underscores the importance of "the stabilizing effect of a persistent seed bank" on the regional ecology.[11]

One of the best-known adaptations of many Mediterranean plants is their adaptation to fire.[12] Fire, being a natural occurrence in the Mediterranean landscape, is a frequent "disturbance," and a wide variety of plants have adapted to it over thousands of years of evolution. The bark of the cork oak, for example, protects the trees against forest fires by acting as a kind of insulation. Many Mediterranean plants contain seeds that need fire to stimulate germination while others will flower only following a fire. Several trees and shrubs resprout vigorously after a fire, and others will release seeds only if they experience fire. As a result of these adaptations, the majority of the vegetation in Mediterranean regions recovers quickly after fire. In fact, the first few years after a fire are often years of robust plant growth and high productivity. Forest fires rarely, if ever, destroy a landscape in this region because the regrowth of the plants, including trees, is vigorous.

Although conventionally considered overgrazed, the Maghreb actually contains a wide variety of plants that are highly adapted to grazing. Many trees and shrubs in the region are stimulated by and grow more vigorously following grazing. This is true of many common plants in the Maghreb such as *Artemesia herba alba* (sagebrush, known locally as *chich*). The local shrub *Atriplex halimus,* for instance, will thrive only when grazed.[13] This shrub, moreover, boosts its protein content when grazed, thereby making it more attractive to grazing animals.[14] Likewise, the Kermes oak common in the Maghreb also grows much more vigorously after grazing.[15] As a result of thousands of years of coevolution with grazing animals, many of the plants in the Maghreb (and the Middle East in general) are well-adapted to heavy grazing.[16] Tillering, the production of side shoots in grasses and other plants, is stimulated by grazing and often results in more robust plant growth and the production of more seeds. Many grasses, especially perennial grasses, also respond to grazing by increasing their overall growth, elongating their leaves, and increasing their rate of photosynthesis. A great many of the grasses that grow in the Maghreb, in particular the annual grasses, are highly resistant to grazing as well as drought.[17]

Equipped with these multiple adaptations to aridity, drought, fire, and grazing, the plants of the Maghreb are highly resilient to "disturbance." Disturbances such as grazing, and fire in particular, were believed for decades to be unnatural events that upset the "balance of nature." Although many still cling to these ideas, most contemporary ecologists have changed their views about notions of equilibrium and disturbance over the past several decades. The conventional conception of fire or grazing as unnatural and upsetting the natural equilibrium has been discredited. Rather, many such "disturbances" are now largely considered to be natural components of the ecosystem. Landscapes all over the world have been wrongly interpreted to be pristine, natural landscapes when in fact they are simply experiencing a phase in a long trajectory of recovery from a disturbance.[18]

Ecology as a discipline has experienced a paradigm shift away from notions of equilibrium, linear succession, and climax communities.[19] Ecological studies of arid lands, and grazing lands in particular, have demonstrated that grazing in many environments, especially in arid lands, does not necessarily have the negative, disturbing effect assumed in conventional, successional ecological theory. In this older model, the ecological system is believed to be maintained by feedback mechanisms, and grazing is thought to nearly always result in a retrogression of the stages of vegetation succession away from the desired perennial and, ideally, arboreal vegetation.[20] This conventional theory of ecological succession has been widely criticized for being overly simplistic, especially with respect to grazing, and has been largely undermined in arid and semiarid regions. Pastoral development projects that have incorporated the older successional model have nearly universally been judged as failures.[21] Newer, less deterministic models have been widely accepted that posit multiple stable states, for instance, rather than a single ideal vegetation climax for a given environment.[22] Ecologists have demonstrated in the Mediterranean Middle East that many of the annual grasslands of the region, conventionally cited as evidence of overgrazing, actually do not represent the degraded state assumed by successional ecology. Rather, they represent healthy vegetation communities.[23] These experiments, which fenced out all animals, showed that grasses formed the "climax" vegetation and that grazing was not, as usually assumed, preventing the growth of shrubby and arboreal vegetation.[24]

As a result of these contemporary ecological theories and increasing evidence of the resilience of arid lands and Mediterranean vegetation, new management strategies have been suggested for these regions that are both sustainable and productive. The use of fire and sometimes grazing as landscape management tools, for example, is now widely accepted. Another large and growing body of research points to the imperative of mobility of livestock herds in arid environments, and to the appropriateness of the common (collective) property management systems of many indigenous pastoralists who understand the crucial timing of herd movements and other ways of managing their environment successfully.[25] The general type of management strategy emerging from this research is one of "opportunistic management," which is often "strikingly similar to the practice[s] of most indigenous African pastoralists."[26] This strategy is also quite similar to the precolonial land management systems in the Maghreb. Thus, nearly two hundred years after their traditional ways of life were condemned and curtailed under colonialism, North African pastoralists and agropastoralists may finally be exonerated.

Notes

Chapter 1

1. Some of the best known include Charles-Robert Ageron, *Les Algériens musulmans et la France (1871–1919)*, 2 vols. (Paris: Presses Universitaires de France, 1968); Yves Lacoste, André Nouschi, and André Prenant, *L'Algérie: Passé et présent* (Paris: Éditions Sociales, 1960); Charles-André Julien, *Histoire de l'Algérie contemporaine: La Conquête et les débuts de la colonisation (1827–1871)* (Paris: Presses Universitaires de France, 1964); Daniel Rivet, *Lyautey et l'instauration du protectorat français au Maroc 1912–1925*, 3 vols. (Paris: Harmattan, 1988); and Jean Poncet and André Raymond, *La Tunisie* (Paris: Presses Universitaires de France, 1971).

2. The call to tell "stories about stories about nature" in environmental history was voiced over a decade ago by William Cronon. See "A Place for Stories: Nature, History and Narrative," *Journal of American History* 78, no. 4 (March 1992): 1347–76.

3. The term "Maghreb" is used throughout this book to refer to the three North African countries of Algeria, Morocco, and Tunisia. It does not include Libya, although other writers sometimes include Libya as part of the Maghreb.

4. P. Christian, *L'Afrique française, l'empire de Maroc et les déserts de Sahara* (Paris: A. Barbier, 1846), 315. All translations from the French are mine unless otherwise noted.

5. Résidence Générale de France à Tunis, Service des Affaires Indigènes, *Historique de l'annexe des affaires indigènes de Ben-Gardane* (Bourg: Victor Berthod, 1931), 13.

6. Augustin Bernard and Napoléon Lacroix, *L'Évolution du nomadisme en Algérie* (Alger: Adolphe Jourdan, 1906), 24.

7. See Paul Boudy, *Économie forestière nord-africaine: Milieu physique et milieu humain*, vol. 1 (Paris: Larose, 1948), 227.

8. Bernard and Lacroix, *L'Évolution*, 24. For a fuller discussion of Ibn Khaldoun and the problematic and selective use of his writings during the colonial period, including the intense debates and controversy among French scholars in the postcolonial period, see chap. 3 below.

9. André Nouschi, *Enquête sur le niveau de vie des populations rurales constantinoises de la conquête jusqu'en 1919* (Paris: Presses Universitaires de France, 1961), 161.

10. Louis Lavauden, "Les Forêts du Sahara," *Revue des Eaux et Forêts* 65, no. 6 (1927): 265–77 and 329–41.

11. Bernard and Lacroix, *L'Évolution*, 42–43.

12. José Germain and Stéphane Faye, *Le Nouveau monde français: Maroc—Algérie—Tunisie* (Paris: Plon, 1924), ii–iii.

13. Brent Shaw, "Climate, Environment, and History: The Case of Roman North Africa," in *Climate and History: Studies in Past Climates and Their Impact on Man*, ed. T. Wigley, M. Ingram, and G. Farmer (Cambridge: Cambridge University Press, 1981), 379–403.

14. Ibid., 389. Reliable estimates of production and domestic consumption of grains in North Africa at this time are not available, to the best of my knowledge.

15. Anon., *Statistique et documents relatifs au Sénatus-Consulte sur la propriété arabe* (Paris: Imprimerie Impériale, 1863), 478.

16. Jean Saint-Germes, *Économie algérienne* (Alger: La Maison des Livres, 1955), 104–5.

17. Annual harvests fluctuate wildly depending on the amount of precipitation in any given year. The cereal harvest in Morocco, for example, was 1,783,230 metric tons in 1995 (a severe drought year) and 10,103,620 metric tons, nearly ten times higher, in 1996 (a year of very good rainfall).

18. Bruno Messerli and Matthias Winiger, "Climate, Environmental Change, and Resources of the African Mountains from the Mediterranean to the Equator," *Mountain Research and Development* 12, no. 4 (1992): 332. See also Jean-Louis Ballais, "Conquests and Land Degradation in the Eastern Maghreb during Classical Antiquity and the Middle Ages," in *The Archaeology of Drylands*, ed. Graeme Barker and David Gilbertson (London: Routledge, 2000), 125–36.

19. Messerli and Winiger, "Climate," 333. Overgrazing has long been blamed for reducing vegetation cover and thus causing significant erosion throughout the Mediterranean basin. Recent research demonstrates, however, that much erosion in the region is instead produced by very strong storms which in short periods of time generate large amounts of rainfall. In many areas,

these storms likely have had a more detrimental impact on the environment than pastoralism. See Karl W. Butzer, "Environmental History in the Mediterranean World: Cross-Disciplinary Investigation of Cause-and-Effect for Degradation and Soil Erosion," *Journal of Archaeological Science* 32, no. 12 (2005): 1773–1800. This is not to say that overgrazing is not a problem at all in the Mediterranean basin, and especially in the Maghreb. It presents a problem, however, in relatively few localized areas where large numbers of livestock are kept in circumscribed areas, especially in commercial operations near cities or villages.

20. The French, of course, were not unique in developing disparaging environmental narratives vilifying indigenous peoples for colonial gain. Similar stories were used in much the same way in British African possessions, in the American West, and elsewhere. See, for example, Roderick P. Neumann, *Imposing Wilderness: Struggles over Livelihood and Nature Preservation in Africa* (Berkeley: University of California Press, 1998); and Karl Jacoby, *Crimes against Nature: Squatters, Poachers, Thieves, and the Hidden History of American Conservation* (Berkeley: University of California Press, 2001).

21. See J.-A.-N. Périer, *De l'hygiène en Algérie*, vol. 1 in the subseries Sciences médicales of the series Exploration scientifique de l'Algérie (Paris: Imprimerie Royale, 1847); and Henri Lorin, *L'Afrique du nord: Tunisie, Algérie, Maroc* (Paris: Armand Colin, 1908).

22. Jean Colin, *L'Occupation romaine du Maroc, cours préparatoire au service des affaires indigènes* (Rabat: Direction Générale des Affaires Indigènes, 1925), 3.

23. Ibid., 13.

24. *Bulletin de la Ligue du Reboisement de l'Algérie,* no. 1, 1882, 2, 6.

25. Clément Alzonne, *L'Algérie* (Paris: Fernand Nathan, 1937), 25 (emphasis mine).

26. Henri Marc, *Notes sur les forêts de l'Algérie,* Collection du centenaire de l'Algérie, 1830–1930 (Paris: Larose, 1930), 497. Batna is located approximately halfway between Constantine and Biskra (see map 1).

27. Christian, *L'Afrique Française,* 52–53. See also Charles Rivière and H. Lecq, *Traité pratique d'agriculture pour le nord de l'Afrique* (Paris: Société d'Éditions, 1928).

28. Maurice Besnier, "La Géographie économique du Maroc dans l'antiquité," *Archives Marocaines* 7, no. 1 (1906): 273.

29. J. Reynard, *Restauration des forêts et des pâturages du sud de l'Algérie (province d'Alger)* (Alger: Adolphe Jourdan, 1880), 6.

30. Boudy, *Économie forestière,* vol. 1, is the authoritative source for most of these estimates.

31. For an important recent discussion of some new interpretations of the northern Mediterranean basin as being far from a deforested, overgrazed, "ruined landscape," see A. T. Grove and Oliver Rackham, *The Nature of Mediterranean Europe: An Ecological History* (New Haven, Conn.: Yale University Press, 2001).

32. See for example, William C. Brice, ed., *Environmental History of the Near and Middle East since the Last Ice Age* (London: Academic Press, 1978); H. Elenga et al., "Pollen-Based Biome Reconstruction for Southern Europe and Africa 18,000 yr BP," *Journal of Biogeography* 27, no. 3 (2000): 11–34; Fernando Lopez-Vera, ed. *Quaternary Climate in Western Mediterranean* (Madrid: Universidad Autónoma de Madrid, 1986); P. Rognon, "Late Quaternary Climatic Reconstruction for the Maghreb (North Africa)," *Palaeogeography, Palaeoclimatology, Palaeoecology* 58, no. 1–2 (1987): 11–34; and Annik Brun, "Microflores et paléovégétations en Afrique du nord depuis 30,000 ans," *Bulletin de la Société Géologique de France* 8, no. 1 (1989): 25–33.

33. Neil Roberts, *The Holocene: An Environmental History* (Oxford: Blackwell, 1998), 63; my calculations are based on the author's table 3.3.

34. See Mark A. Blumler, "Biogeography of Land-Use Impacts in the Near East," in *Nature's Geography: New Lessons for Conservation in Developing Countries,* ed. Karl Zimmerer and Kenneth Young (Madison: University of Wisconsin Press, 1998), 215–36; and Luc Wengler and Jean-Louis Vernet, "Vegetation, Sedimentary Deposits and Climates during the Late Pleistocene and Holocene in Eastern Morocco," *Palaeogeography, Palaeoclimatology, Palaeoecology* 94, no. 1–4 (1992): 141–67.

35. Rognon, "Late Quaternary"; and Wengler and Vernet, "Vegetation."

36. Messerli and Winiger, "Climate." Most of these tropical mammals, such as the rhinoceros and the hippopotamus, probably became extinct in North Africa well before the Roman period or survived only in very small numbers in areas with adequate water sources. Many large mammals were collected in North Africa to provide entertainment in Roman arenas. Although this undoubtedly did decrease the population of many species, the extinction of most of these animals was more likely due to habitat modification and the introduction of firearms in the last half of the nineteenth century. See Brent Shaw, "Climate," 385–87.

37. James C. Ritchie, "Analyse pollinique de sédiments holocènes supérieurs des hauts plateaux du Maghreb oriental," *Pollen et Spores* 26, no. 3–4 (1984): 489–96.

38. Aziz Ballouche et al., "Holocene Environments of Coastal and Continental Morocco," in *Quaternary Climate in Western Mediterranean,* ed. Fernando

Lopez-Vera (Madrid: Univ. Autónoma, 1986), 517–31; H. F. Lamb, U. Eichner, and V. R. Switsur, "An 18,000-Year Record of Vegetation, Lake-Level and Climate Change from Tigalmamine, Middle Atlas, Morocco," *Journal of Biogeography* 1, no. 16 (1989): 65–74; H. F. Lamb et al., "Lacustrine Sedimentation in a High-Altitude, Semi-arid Environment: The Palaeolimnological Record of Lake Isli, High Atlas, Morocco," in *Environmental Change in Drylands: Biogeographical and Geomorphological Perspectives,* ed. Andrew C. Millington and Ken Pye (New York: John Wiley, 1994), 147–61; and Ritchie, "Analyse Pollinique."

39. Ritchie, "Analyse Pollinique"; Rognon, "Late Quaternary"; Annik Brun, "Étude palynologique des sédiments marins holocènes de 5000 B.P. à l'actuel dans le golfe de Gabès," *Pollen et Spores* 25, no. 3–4 (1983): 437–60; D. Faust et al., "High-Resolution Fluvial Record of Late Holocene Geomorphic Change in Northern Tunisia: Climate or Human Impact?" *Quaternary Science Reviews* 23, no. 16–17 (2004): 1757–75.

40. H. F. Lamb, F. Damblon, and R. W. Maxted, "Human Impact on the Vegetation of the Middle Atlas, Morocco, during the Last 5000 Years," *Journal of Biogeography* 18, no. 5 (1991): 519–32.

41. Lamb et al., "Lacustrine Sedimentation."

42. Maurice Reille, "Contribution pollenanalytique à l'histoire holocène de la végétation des montagnes du Rif (Maroc septentrional)," *Supplément au Bulletin AFEQ* 1, no. 50 (1977): 53–76; and J. Bernard and M. Reille, "Nouvelles analyses polliniques dans l'Atlas de Marrakech, Maroc," *Pollen et Spores* 29, no. 2–3 (1987): 225–40.

An exception to this general pattern may be the Rif Mountains in the north of Morocco. Reille's (1977) article on the Rif Mountains shows greater decline in more species than other fossil pollen studies in Morocco. One of the first pollen studies in Morocco, it has probably received the most scholarly attention and the highest number of citations to date. His study is limited, however, because it covered no more than a seventy-five kilometer region of the Rif Mountains and it was not carbon dated. In fact, it appears that his dates are mostly inferred from the standard environmental history described in this book. He concluded that the "Arab invasion [of the tenth century] has been the indirect cause of an intensive deforestation." Reille, "Contribution," 53. More recent research based on tree-ring analysis in the Rif, Middle Atlas, and High Atlas has revealed that Morocco suffered a severe period of drought from roughly 1100 to 1250 CE. See Claudine Till and Joel Guiot, "Reconstruction of Precipitation in Morocco since 100 A.D. Based on *Cedrus atlantica* Tree-Ring Widths," *Quaternary Research* 33, no. 3 (1990):

337–51. It may be this period of significant drought that Reille's pollen cores are showing, rather than deforestation from the "Arab invasion." Several of his twelve cores actually show recent (though in an unknown time period) *increases* in many tree species. Likewise, the article by Bernard and Reille (1987) on the High Atlas shows more deforestation than other studies, but it also lacks carbon dating. It is interesting to compare the Bernard and Reille article—which blames deforestation on human impact, especially the "Arab invasion"—with Lamb et al., "Lacustrine Sedimentation" (1994), based on samples taken from a lake in the High Atlas only about 150 miles away: they were well carbon-dated and do not show significant deforestation over the last three thousand years. Lamb also tentatively concluded that "it is less likely that anthropogenic influence on soil erosion can be inferred from the sedimentary record" (160).

43. Ritchie, "Analyse Pollinique."

44. Ibid.; M. Salamani, "Premières données paléophytogéographiques du cèdre de l'Atlas (*Cedrus atlanticus*) dans la région de grande Kabylie (N-E Algérie)," *Palynosciences* 2 (1993): 147–55.

45. Blumler, "Biogeography," 220.

46. Michel Thinon, Aziz Ballouche, and Maurice Reille, "Holocene Vegetation of the Central Saharan Mountains: The End of a Myth," *Holocene* 6, no. 4 (1996): 457–62.

47. Ibid., 460. Quote from Reynard, *Restauration*, 6.

48. Brun, "Étude Palynologique."

49. Ibid.; Rognon, "Late Quaternary"; Faust et al., "High-Resolution"; Lamb et al., "Lacustrine Sedimentation"; and Lamb, Eichner, and Switsur, "18,000-Year Record." There is further evidence that aridification is increasing on the African continent, and may have been in a fluctuating pattern for the last five hundred years. See Sharon E. Nicholson, "Climatic Variations in the Sahel and Other African Regions during the Last Five Centuries," *Journal of Arid Environments* 1, no. 11 (1978): 3–24; and "Recent Rainfall Fluctuations in Africa and Their Relationship to Conditions over the Continent," *Holocene* 4, no. 2 (1994): 121–31. See also James L. A. Webb, *Desert Frontier: Ecological and Economic Change along the Western Sahel, 1600–1850* (Madison: University of Wisconsin Press, 1995) for a good discussion on West Africa. Although some of this evidence points to continued aridification over the last two thousand years, the aridification has been gradual and has not been of the magnitude of the earlier major shifts, such as occurred around three to four thousand years before present or at the end of the last Ice Age.

50. B. Shaw, "Climate," 392.

51. Because of the arid conditions in much of North Africa, pollen analysis is not possible in many areas. Reconstruction of climate history and estimates of past vegetation are therefore still hypothetical to varying degrees.

52. It is important to note that those working on the environment had a choice in how they interpreted the vegetation and its past. It would not have been impossible for these men to have made different choices, just as it would have been easy for the French to view the Arabs and the Berbers as related and friendly instead of as two separate "races" that had been at war for centuries. The choices made by ecologists working in the Maghreb are discussed in detail in chap. 5.

53. For two well-known examples, see William Cronon, *Changes in the Land: Indians, Colonists, and the Ecology of New England* (New Haven, Conn.: Yale University Press, 1983); and James Fairhead and Melissa Leach, *Reframing Deforestation: Global Analyses and Local Realities; Studies in West Africa* (London: Routledge, 1998).

54. Stéphane Gsell, *Les conditions du développement historique, les temps primitifs, la colonisation phénicienne et l'empire de Carthage,* vol. 1 of *Histoire ancienne de l'Afrique du nord,* 3rd ed. (Paris: Hachette, 1921), 158.

55. Ibid., 155.

56. Throughout this book, I use the term "Algerian(s)" to refer to the indigenous populations of Algeria. I use the words settler or colonist to refer to the French and European immigrants during the colonial period, despite the fact that many of these European settlers called themselves "Algerians." Similarly, I reserve the terms Moroccan(s) and Tunisian(s) for the indigenous populations.

57. For more detailed information on the physical environment of the region, see the Appendix, "A Note on the Geography and Ecology of the Maghreb." Readers not familiar with the geography of North Africa will find this appendix very helpful in contextualizing the rest of the book.

58. Pastoral nomadism is generally defined as movement of livestock over large areas of relatively flat terrain, such as the Saharan borderlands, during most or all of the year. Another form of migration, transhumance, is the movement of livestock to higher elevations during the warm summers and down to the lowlands during the cool winters. Transhumance was quite common in the mountains of Morocco, for example, and still is in some areas.

Chapter 2

1. For an excellent discussion of pre-nineteenth-century European attitudes toward the Maghreb, see Ann Thomson, *Barbary and Enlightenment:*

European Attitudes towards the Maghreb in the 18th Century (Leiden: E. J. Brill, 1987). Although this book focuses on the eighteenth century, it contains a brief discussion of earlier sources dating back to the seventeenth century. For a discussion of early French travelers to the Maghreb, see Denise Brahimi, *Voyageurs français du XVIIIe siècle en Barbarie* (Lille: Atelier Reproduction des Thèses, 1976).

2. Pliny, *Natural History*, vol. 2, *Books 3–7*, trans. H. Rackham (Cambridge, Mass.: Harvard University Press, 1942), 225.

3. Ibid., 229.

4. Strabo, quoted in P. Christian, *L'Afrique française, l'empire de Maroc et les déserts de Sahara* (Paris: A. Barbier, 1846), 52.

5. The classics were widely read in France in the eighteenth century and had a significant impact on French society. See Harold Parker, *The Cult of Antiquity and the French Revolutionaries* (Chicago: University of Chicago Press, 1937). It should be noted that French (and other) authors borrowed liberally from ancient sources, often without attribution or citation.

6. Thomas Shaw, *Travels or Observations Relating to Several Parts of Barbary and the Levant* (Oxford: Printed at the Theatre, 1738).

7. Shaw was very familiar with the ancient writers on North Africa and he deliberately patterned himself on several of the ancient geographers. "Strabo, Ptolemy, and Pliny," he explained, "those celebrated masters in this branch of literature, have given us the pattern, which we have all along endeavoured to follow and imitate." Thomas Shaw, "Travels or Observations Relating to Barbary," in *A General Collection of the Best and Most Interesting Voyages and Travels in All Parts of the World*, ed. John Pinkerton (London: Longman, Hurst, Rees, Orme, and Brown, 1814), 513. This 1814 printing in Pinkerton's series is a faithful copy of the 1757 second edition of Shaw's travels, though it includes only the sections on Barbary and not on the Levant. Subsequent citations of Shaw refer to this edition. Shaw also took some of these ancient authors to task for their exaggerations and fanciful imaginings (see, e.g., 512).

8. William Shaler, *Sketches of Algiers, Political, Historical and Civil, Containing an Account of the Geography, Population, Government, Revenues, Commerce, Agriculture, Arts, Civil Institutions, Tribes, Manners, Language and Recent Political History of That Country* (Boston: Cummings, Hilliard, 1826).

9. G. T. Raynal, *Histoire philosophique et politique des établissements et du commerce des Européens dans l'Afrique septentrionale*, 2 vols. (Paris: Amable Costes, 1826). Raynal's work was published posthumously from his notes. For a discussion of the major written sources on Algeria used by the French around the time of conquest, see Patricia M. Lorcin, "Rome and France in

Africa: Recovering Colonial Algeria's Latin Past," *French Historical Studies* 25, no. 2 (2002): 298.

10. Jean-André Peyssonnel, "Relation d'un voyage sur les côtes de la Barbarie fait par ordre du roi en 1724 et 1725," in *Voyages dans les régences de Tunis et d'Alger*, vol. 1, ed. M. Dureau de la Malle (Paris: Gide, 1838).

11. René Louiche Desfontaines, "Fragmens d'un voyage dans les régences de Tunis et d'Alger, fait de 1785 à 1786," in *Voyages dans les régences de Tunis et d'Alger*, vol. 2, ed. M. Dureau de la Malle (Paris: Gide, 1838).

12. T. Shaw, "Travels," 523.

13. William Shaler, *Esquisse de l'état d'Alger considéré sous les rapports politique, historique et civil*, trans. Thomas-Xavier Bianchi (Paris: Ladvocat, 1830), 12, 14. Subsequent citations of Shaler refer to this edition.

14. Raynal, *Histoire*, 1:183. Ann Thomson explains that Raynal's work was not written by him alone but that numerous other authors contributed, including Diderot. See Thomson, *Barbary*, 147n9.

15. Raynal, *Histoire*, 1:182.

16. Peyssonnel, "Relation," 276, 280, 330.

17. Desfontaines, "Fragmens," 260. See also 16, 138–94.

18. Ibid., 165.

19. T. Shaw, "Travels," 523.

20. M. Renaudot, *Alger: Tableau du royaume, de la ville d'Alger et de ses environs*, 4th ed. (Paris: P. Mongie Ainé, 1830), 123.

21. Desfontaines, "Fragmens," esp. chaps 5 and 6.

22. Shaw seems to be an exception to this trend. He is quite disparaging, though, of many of the fruits and vegetables he finds in North Africa, whose taste he does not appreciate as much as that of European varieties.

23. Shaler, *Esquisse*, 22. See also 112.

24. Desfontaines, "Fragmens," 271.

25. Raynal, *Histoire*, 2:141. Raynal is quoting (in agreement) an Italian author in this passage, which continues, "scarcely does the plough leave a trace on the fields . . . the pastures are truly well watered; but the inhabitants absolutely ignore the art of raising and nourishing the sheep and the horned animals."

26. Renaudot, *Alger*, 128.

27. Ibid., 5.

28. See Thomson, *Barbary*, 64–93, for a much more detailed analysis of the extremely complex and contrasting population classifications attempted by many of the authors on North Africa during this period.

29. See Raynal, *Histoire*, 2:143–55; and Renaudot, *Alger*, 23–55, for their descriptions of the local populations.

30. Raynal was interested in the blacks, many of whom were slaves, particularly because he was opposed to slavery, as were most of the religious authors of the period.

31. Renaudot, *Alger,* 131.

32. Raynal wrote that "the Arabs work their fields: they plant wheat, and plant beans towards mid-October; barley, lentils and other vegetables are not planted until 15 days later. These diverse cultivations are not very efficient; but the soil there is so fertile that the productions arrive in abundance." Raynal, *Histoire,* 1:185.

33. Renaudot, *Alger,* 128.

34. Shaler, *Esquisse,* 111–12.

35. Raynal, *Histoire,* 2:149–50.

36. See the discussion of the complicated idealization and denigration of the bedouin, the Berbers, and other North Africans in Thomson, *Barbary,* 76–90, 104–18.

37. Desfontaines, "Fragmens," 20.

38. Peyssonnel, "Relation," 73, 222 (emphasis in original).

39. Raynal, *Histoire,* 2:152–53.

40. Ibid., 2:153.

41. Renaudot, *Alger,* 38; T. Shaw, "Travels," 506. Contrast this, however, with Shaw's negative description of the pillaging "wild Arabs," 505.

42. Peyssonnel, "Relation," 206.

43. Raynal, *Histoire,* 1:23.

44. See Thomson, *Barbary;* and Patricia M. Lorcin, *Imperial Identities: Stereotyping, Prejudice and Race in Colonial Algeria* (London: I. B. Tauris, 1995), 19–21, 32.

45. Peyssonnel, "Relation," 209.

46. Desfontaines, "Fragmens," 6–7.

47. Renaudot, *Alger,* 181.

48. Ibid., 181–82. French scholars have argued, however, that the extent and importance of piracy have been exaggerated. See Lucette Valensi, *On the Eve of Colonialism: North Africa before the French Conquest,* trans. Kenneth J. Perkins (New York: Africana, 1977).

49. Mahfoud Bennoune, *The Making of Contemporary Algeria, 1830–1987: Colonial Upheavals and Post-independence Development* (Cambridge: Cambridge University Press, 1988), 29–31. As early as 1560, merchants in Marseilles obtained special permission from the government of Algiers to gather coral off the coast and in 1694 the French Africa Company benefited from a convention signed between the French king and the dey of Algiers regarding com-

mercial relations (including the importation of wheat to France) and protection from pirates. See Raynal, *Histoire*, 1:194–95.

50. King Charles X had come to power in 1824, following a decade of rule by his brother, Louis XVIII, who was placed on the throne in 1815 when the Bourbon monarchy was restored. Their brother, Louis XVI, had been executed in 1793 in the bloody aftermath of the French Revolution of 1789. For a good survey of French history during this period, see Roger Price, *A Concise History of France* (Cambridge: Cambridge University Press, 1993).

51. For details of these events, see Bennoune, *Making,* and Charles-André Julien, *Histoire de l'Algérie contemporaine: La Conquête et les débuts de la colonisation (1827–1871)* (Paris: Presses Universitaires de France, 1964).

52. John Ruedy, *Modern Algeria: The Origins and Development of a Nation* (Bloomington: Indiana University Press, 1992), 46.

53. For various renditions of this legendary incident and the history behind it, see Julien, *Histoire;* Bennoune, *Making;* and Ruedy, *Modern.*

54. For details see Robert Aldrich, *Greater France: A History of French Overseas Expansion* (London: Macmillan, 1996). A few of the earlier territories were restored to France before the conquest of Algeria.

55. Prosper Enfantin, *Colonisation de l'Algérie* (Paris: P. Bertrand, 1843), 31.

56. Ibid., 31–32.

57. From the report of the African Commission, quoted in Bennoune, *Making,* 35.

58. Ibid.

59. See Julien, *Histoire;* and Yves Lacoste, André Nouschi, and André Prenant, *L'Algérie: Passé et présent* (Paris: Éditions Sociales, 1960).

60. Ruedy, *Modern,* 23.

61. Lacoste, Nouschi, and Prenant, *L'Algérie,* 189–90.

62. For statistical details and more information, see André Nouschi, "Notes sur la vie traditionelle des populations forestières algériennes," *Annales de Géographie* 68, no. 370 (1959): 525–35.

63. Lacoste, Nouschi, and Prenant, *L'Algérie,* 217–25; Julien, *Histoire,* 7.

64. M'hammed Boukhobza, *Monde rural: Contraintes et mutations* (Alger: Office des Publications Universitaires, 1992), 26.

65. This practice is called *écobuage* in French and is described by Julien, *Histoire,* 7–8. See also Renaudot, *Alger,* 131. Livestock sometimes grazed the stubble after the harvests, which added rich manure to the soil.

66. For details on *ksir* see Nouschi, "Notes," 531; André Nouschi, *Enquête sur le niveau de vie des populations rurales constantinoises de la conquête jusqu'en 1919* (Paris: Presses Universitaires de France, 1961); and Charles-Robert Ageron,

Les Algériens musulmans et la France (1871–1919) (Paris: Presses Universitaires de France, 1968), 1:105.

67. Karen Pfeifer, "The Development of Commercial Agriculture in Algeria, 1830–1970," *Research in Economic History* 10 (1986): 277–308.

68. Despite the simplistic French colonial categorization of "Arab nomads" and "sedentary Berbers," many Berbers in North Africa were indeed nomads and seminomads, just as many Arabs were village-based and not nomadic.

69. Julien, *Histoire,* 8.

70. A similar kind of biannual migration of seminomads and their livestock took place in many mountainous areas. Called transhumance, it meant summers were spent in alpine pastures and winters were passed in lowlands and valleys.

71. Lacoste, Nouschi, and Prenant, *L'Algérie,* 226.

72. The subject of property and land tenure in precolonial Algeria is controversial. For details and good discussions, see Ageron, *Les Algériens,* 67–78; Bennoune, *Making,* 24–26, esp. the quote on 26; and Lacoste, Nouschi, and Prenant, *L'Algérie,* 205–10.

73. Ageron, *Les Algériens,* 67–78, contains an especially detailed and illuminating discussion of the evolution of the French understanding of property rights in Algeria during the early stages of the occupation.

74. For an excellent discussion of arid lands ecology and pastoral livelihoods, see Ian Scoones, ed., *Living with Uncertainty: New Directions in Pastoral Development in Africa* (London: Intermediate Technology Publications, 1995).

75. Mark A. Blumler, "Biogeography of Land-Use Impacts in the Near East," in *Nature's Geography: New Lessons for Conservation in Developing Countries,* ed. Karl Zimmerer and Kenneth Young (Madison: University of Wisconsin Press, 1998), 215–36. See also Avi Perevolotsky and No'am Seligman, "Role of Grazing in Mediterranean Rangeland Ecosystems," *BioScience* 48, no. 12 (1998): 1007–17; and Michael B. Coughenour, "Graminoid Responses to Grazing by Large Herbivores: Adaptations, Exaptations, and Interacting Processes," *Annals of the Missouri Botanical Garden* 72, no. 4 (1985): 852–63.

76. Louis Trabaud, "Postfire Plant Community Dynamics in the Mediterranean Basin," in *The Role of Fire in Mediterranean-Type Ecosystems,* ed. José Moreno and Walter Oechel (New York: Springer-Verlag, 1994), 1–15.

77. A. T. Grove and Oliver Rackham, *The Nature of Mediterranean Europe: An Ecological History* (New Haven, Conn.: Yale University Press, 2001), 218; and Zev Naveh, "The Role of Fire and Its Management in the Conservation of Mediterranean Ecosystems and Landscapes," in *The Role of Fire in*

Mediterranean-Type Ecosystems, ed. José Moreno and Walter Oechel (New York: Springer-Verlag, 1994), 163–85.

78. Nouschi, "Notes," 531; and Grove and Rackham, *Nature,* 218.

79. King Louis-Philippe signed a royal ordinance on 22 July 1834 that formalized the existence of "the French possessions of North Africa." Julien, *Histoire,* 114.

80. For a discussion of some of the pro- and anticolonization arguments and actors, see ibid., 106–63.

81. Bugeaud's method of pacification has been described as a "scorched earth" policy in which he and other military leaders routinely burned fields, cut down orchards, and slaughtered livestock as methods of "pacification." For details, see Bennoune, *Making,* 39–41; Julien, *Histoire,* chap. 3; and Charles-Robert Ageron, *Modern Algeria: A History from 1830 to the Present,* trans. Michael Brett (London: Hearst, 1991), 18–21.

82. Nouschi, *Enquête,* 161 (emphasis mine).

83. Ibid.

84. Ruedy, *Modern,* 69.

85. Centre des Archives d'Outre-Mer (CAOM), ALG, GGA, 8H/10, letter from Capitaine Lapasset of the Arab Bureaux to the duc d'Aumale, governor-general of Algeria, 12 January 1848.

86. Henri Marc, *Notes sur les forêts de l'Algérie* (Alger: Adolphe Jourdan, 1916), 208.

87. Ageron, *Les Algériens,* 107n4. See also Achilles Filias, *Notice sur les forêts de l'Algérie: Leur étendue; leurs essences; leurs produits* (Alger: Gojosso, 1878), 4.

88. Ageron, *Les Algériens,* 1:107.

89. Paul Boudy, *Économie forestière nord-africaine: Milieu physique et milieu humain,* (Paris: Larose, 1948), 1:230–31.

90. Ibid., 231.

91. Ibid.

92. Ibid.

93. Ageron, *Modern,* 24.

94. Filias, *Notice,* 3.

95. Field Marshal de Saint-Arnaud, quoted in Anne Bergeret, "Discours et politique forestières coloniales en Afrique et à Madagascar," *Revue Française d'Histoire d'Outre-Mer* 79, no. 298 (1993): 24–25.

96. Boudy, *Économie forestière,* 1:233.

97. See Angel-Paul Carayol, *La Législation forestière de l'Algérie* (Paris: Arthur Rousseau, 1906), 24.

98. Ageron, *Les Algériens,* 1:106. See also Boudy, *Économie forestière,* 1:233.

99. Marc, *Notes,* 57.

100. Renou was trained at the National Forestry School at Nancy, from which he graduated in 1830. Boudy, *Économie forestière,* 1:230. In addition to being an excellent forester he was known as a good botanist. René Maire, *Les Progrés des connaissances botaniques en Algérie depuis 1830* (Paris: Masson, 1931), 23.

101. This article, "Notes sur les Forêts de Cèdres de l'Algérie" appeared in the *Annales Forestières* 3 (1844): 1–7, and is reprinted in Maire, *Les Progrés,* 25–36.

102. Munby described the excellent natural pastures around Algiers and the rich variety of shrubs and trees in the area. He explained the traditional practice of leaving stubble in the cereal fields after harvest for animal graz- ing, as well as the regular burning of this stubble prior to planting, noting that "the ashes serve to fertilize the earth for another harvest." G. Munby, *Flore de l'Algérie ou catalogue des plantes indigènes du royaume d'Alger* (Paris: J.-B. Baillière, 1847), xi.

103. Munby lived in Algeria from 1839 to 1847. The only other important amateur botanist of the period was Ad Steinheil, whose work appeared dur- ing the 1830s. This author, however, limited himself to plant morphology and musings on the global distributions of certain plants found in Algeria, with- out mention of the overall state of the vegetation or any consideration of degradation. See Ad Steinheil, "Matériaux pour servir à la flore de Barbarie," *Annales des Sciences Naturelles* 9 (1838): 193–211.

104. For an excellent discussion of the background to and activities of the commission, see Michael J. Heffernan, "An Imperial Utopia: French Surveys of North Africa in the Early Colonial Period," in *Maps and Africa,* ed. J. Stone (Aberdeen: Aberdeen University African Studies Group, 1994), 81–107. The Scientific Commission of Algeria had much in common with, and was partly modeled on, the *Description de l'Egypte.* For an informative discussion of this earlier French venture, the resulting ten volumes, and the uses to which they were put, see Anne Godlewska, "Map, Text and Image. The Mentality of En- lightened Conquerors: A New Look at the *Description de l'Egypte,*" *Transac- tions of the Institute of British Geographers* 20, no. 1 (1995): 5–28.

105. Some of the last research was carried out in the 1850s and 1860s. The most important volume on botany, for example, appeared in 1867 after field work had been carried out from 1852 to 1861 by M. E. Cosson. See Ernest Cos- son and M. Durieu de Maisonneuve, *Flore d'Algérie: Phanérogamie; Groupe des glumacées,* vol. 2 in the subseries Sciences naturelles: Botanique of the series Exploration scientifique de l'Algérie (Paris: Imprimerie Impériale, 1855–67).

106. Heffernan, "Imperial Utopia," 81.

107. Carette wrote the first five volumes of the sixteen-volume section on the historical and geographical sciences and served as secretary of the Commission Scientifique.

108. Ernest Carette, *Études sur la Kabilie,* vol. 4 in the subseries Sciences historiques et géographiques of the series Exploration scientifique de l'Algérie (Paris: Imprimerie Nationale, 1848), 245.

109. E. Pellissier de Reynaud, *Mémoires historiques et géographiques sur l'Algérie,* vol. 6 in the subseries Sciences historiques et géographiques of the series Exploration scientifique de l'Algérie (Paris: Imprimerie Royale, 1844), 305–6.

110. Archeologists, however, have argued that the existence of aqueducts and water storage devices like cisterns instead indicates that water was as rare and precious during the Roman period as it was under French occupation. This was even recognized by a few specialists during the colonial period. See, for example, Gaston Boissier, *L'Afrique romaine: Promenades archéologiques en Algérie et en Tunisie* (Paris: Hachette, 1895), esp. 136–40.

111. Pellissier de Reynaud, *Exploration,* 307.

112. The primary subject of Périer's treatise was "hygiene" for the successful occupation and colonization of the colony. The intended readers were military doctors and others in the military on duty in Algeria, although Périer stated that the information applied equally well to businessmen and colonists. J.-A.-N. Périer, *De l'hygiène en Algérie,* vol. 1 in the subseries Sciences médicales of the series Exploration scientifique de l'Algérie (Paris: Imprimerie Royale, 1847, v.

113. Ibid., 29.

114. Ibid., 10.

115. Ibid., 30.

116. Lorcin, "Rome and France," 327.

117. See Parker, *Cult of Antiquity.* Moreover, the conquest of Italy and Greece by the French in the late eighteenth and early nineteenth centuries stimulated a veritable pilgrimage of scientists and scholars determined to catalog their ancient treasures.

118. For a good discussion of this early nineteenth-century phenomenon, see Heffernan, "Imperial Utopia," 82–83.

119. Edouard Lapène, *Vingt-six mois à Bougie ou collection de mémoires sur sa conquête, son occupation et son avenir* (Paris: Bouchene, 2002), 131. The experience of the Romans in North Africa also served as a model for the French military during the years of occupation and territorial expansion. General

Bugeaud, for example, studied and implemented Roman warfare tactics, such as mobile columns, which were new to the French army. Jacques Frémeaux, "Souvenirs de Rome et présence française au Maghreb: Essai d'investigation," in *Connaissances du Maghreb: Sciences sociales et colonisation,* ed. Jean-Claude Vatin (Paris: CNRS, 1984), 29–46, 38; and Lorcin, "Rome and France," 299.

120. Frémeaux, "Souvenirs," 32.

121. Ibid., 32–36.

122. Périer, *De l'hygiène en Algérie,* 13.

123. Louis Moll, *Colonisation et agriculture de l'Algérie* (Paris: Librairie Agricole de la Maison Rustique, 1845), 85–86.

124. Ibid., 108.

125. Ibid. This is a particularly early manifestation of what later became known as "social Darwinism," decades before Darwin's ideas were published.

126. Julien, *Histoire,* 215–16. See also Ageron, *Modern,* 23.

127. See Ruedy, *Modern,* 73; and Julien, *Histoire,* 216–18.

128. Bennoune, *Making,* 43. See also Ageron, *Modern,* 25–26.

129. Julien, *Histoire,* 241.

130. Ibid., 240–41. The 1844 law was similar, in some ways, to the French law of 1860 on uncultivated lands and the reclamation of marshes. The French law had an effect comparable to the Algerian law, and "it represented no less than a full assault on the communal lands that the state deemed unproductive." Tamara L. Whited, *Forests and Peasant Politics in Modern France* (New Haven, Conn.: Yale University Press, 2000), 61. The similarities raise the question of whether the Algerian law provided an example that was followed in at least some aspects of the 1860 French law.

131. Julien, *Histoire,* 241.

132. Quoted in Lacoste, Nouschi, and Prenant, *L'Algérie,* 202.

133. See Ernest Carette and Auguste Warnier, *Description et division de l'Algérie* (Paris: L. Hachette, 1847), esp. 22–26.

134. Adolphe Dureau de la Malle, *Province de Constantine: Recueil de rensseignemens pour l'expédition ou l'établissement des français dans cette partie de l'Afrique septentrionale* (Paris: Gide, 1837), esp. xi–xiii, 77.

135. Comte de Raousset-Boulbon, *De la colonisation et des institutions civiles en Algérie* (Paris: Dauvin et Fontaine, 1847), 21.

136. Ibid., 25, 44.

137. M. Delpech de Saint-Guilhem et al., *Mémoire au roi et aux chambres, par les colons de l'Algérie* (Paris: Rignoux, 1847), 10.

138. Heffernan, "Imperial Utopia," 83.

139. Amédée Desjobert, *L'Algérie en 1846* (Paris: Guillaumin, 1846), 6.

140. Ibid., 5.

141. Amédée Desjobert, *L'Algérie* (Paris: Guillaumin, 1847), 36–38.

142. Ibid., 11.

143. Ibid., 9. Interestingly, Desjobert cited Moll's agricultural treatise and Périer's contribution in the *Exploration Scientifique.* He did not, however, cite the section of Périer's work that detailed the declensionist colonial environmental narrative.

144. For more details, see Michael A. Osborne, *Nature, the Exotic, and the Science of French Colonialism* (Bloomington: Indiana University Press, 1994), chap. 6.

145. Auguste Hardy, "Note climatologique sur l'Algérie au point de vue agricole," in *Recueil de traités d'agriculture et d'hygiène à l'usage des colons de l'Algérie,* ed. Ministre de la Guerre (Alger: Imprimerie du Gouvernement, 1851), 37–62, 47.

146. Ibid., 48.

147. Ibid., 55.

Chapter 3

1. Assimilation, according to the settlers' point of view, meant the application of French laws, economic policies, and norms (especially those benefiting the colonists in terms of land acquisition, etc.), and, in effect, destroying Muslim institutions. Thus, for example, several attempts were made to place the Algerians completely under French law and deny them access to Muslim law. As assimilation was meant to eradicate any feelings of Muslim Algerian national identity, this became one of the primary goals of the settlers and their supporters. See Charles-Robert Ageron, *Modern Algeria: A History from 1830 to the Present,* trans. Michael Brett (London: Hearst, 1991), 24, 27, 34–36; and Robert Aldrich, *Greater France: A History of French Overseas Expansion* (London: Macmillan, 1996), 110. See also Lorcin, *Imperial Identities: Stereotyping, Prejudice and Race in Colonial Algeria* (London: I. B. Tauris, 1995); and Raymond Betts, *Assimilation and Association in French Colonial Theory, 1890–1914* (New York: Columbia University Press, 1961).

2. Napoléon III, quoted in Roger Price, *A Concise History of France* (Cambridge: Cambridge University Press, 1993), 179.

3. Ibid., 183.

4. Widespread support for the Empire persisted in the countryside, though, providing it with a majority of votes in the 1870 plebiscite.

5. Ageron, *Modern,* 31.

6. For details on these laws, see Yves Lacoste, André Nouschi, and André Prenant, *L'Algérie: Passé et présent* (Paris: Éditions Sociales, 1960), 361–63; and Charles-André Julien, *Histoire de l'Algérie contemporaine: La Conquête et les débuts de la colonisation (1827–1871)* (Paris: Presses Universitaires de France, 1964), 377–79.

7. Nonagricultural Algerian products also received partial relief from tariffs when exported to France.

8. Lacoste, Nouschi, and Prenant, *L'Algérie,* 361.

9. For details on cantonment, see Lacoste, Nouschi, and Prenant, *L'Algérie,* 363–64; and Ageron, *Modern,* 31–32.

10. Randon, quoted in Julien, *Histoire,* 405.

11. Ibid. Cantonment actually began on a small scale in the 1840s and continued until about 1860. See Julien, *Histoire,* 404–6 and 422, for details. Cantonment had been used in French forests since the 1790s. For a useful discussion of the changing uses of cantonment in France, see Tamara L. Whited, *Forests and Peasant Politics in Modern France* (New Haven, Conn.: Yale University Press, 2000), 35–37.

12. Julien, *Histoire,* 411.

13. See Ageron, *Modern,* part 1; and Julien, *Histoire,* chaps. 7 and 8 for details of this period.

14. Ageron, *Modern,* 35.

15. These men, also called *"Arabophiles"* (Arab lovers), were champions of the indigenous Algerian populations and were often perceived by the settlers as being opposed to settler rights.

16. For more details on Napoléon III's Algerian policies, see Julien, *Histoire,* 419–40; Lacoste, Nouschi, and Prenant, *L'Algérie,* 365–70; and Ageron, *Modern,* 36–44.

17. Michael Heffernan and Keith Sutton, "The Landscape of Colonialism: The Impact of French Colonial Rule on the Algerian Rural Settlement Pattern, 1830–1987," in *Colonialism and Development in the Contemporary World,* ed. Christopher Dixon and Michael Heffernan (London: Mansell, 1991), 130.

18. Napoléon III, quoted in Ageron, *Modern,* 38. *"Colons"* is the French word for colonists or settlers.

19. Ageron, *Modern,* 38.

20. Ibid., 39. This law allowed Muslim Algerians to be employed in military or civilian positions and to apply for full French citizenship.

21. See Julien, *Histoire,* chap. 8 and Ageron, *Modern,* 44–46 for details of this period.

22. See Julien, *Histoire*, chaps. 2–5, 7, and 8 for details of these transformations.

23. Ageron, *Modern*, 23.

24. Ibid., 34, provides just one example. See also John Ruedy, *Modern Algeria: The Origins and Development of a Nation* (Bloomington: Indiana University Press, 1992), 73.

25. This phrase is from Ageron, *Modern*, chap. 4. The details of this period are covered in my next chapter, "The Triumph of the Narrative," the title of which was inspired, in part, by Ageron's work.

26. See Ageron, *Modern*, for example, especially 31–34.

27. Ibid., 33.

28. Heffernan and Sutton, "Landscape of Colonialism," 125. For a more detailed discussion of the influence of the Saint-Simonians in Algeria, see Michael J. Heffernan, "An Imperial Utopia: French Surveys of North Africa in the Early Colonial Period," in *Maps and Africa*, ed. J. Stone (Aberdeen: Aberdeen University African Studies Group, 1994), 81–107; and Lorcin, *Imperial Identities*, chap. 5.

29. Ferdinand Lapasset, *Aperçu sur l'organisation des indigènes dans les territoires militaires et dans les territoires civils* (Alger: Dubos Frères, 1850), 16.

30. Centre des Archives d'Outre-Mer (CAOM), ALG, GGA, 8H/10, letter from Capitaine Lapasset of the Arab Bureaus to the duc d'Aumale, governor-general of Algeria, 12 January 1848. Trying to "fix the Arabs to the soil," that is, to sedentarize the nomads, was already a long-standing goal of the military administration, reaching back to the 1830s with Bugeaud. See Lorcin, *Imperial Identities*, 38–39.

31. See Ageron, *Modern*, 37–44, and Julien, *Histoire*, 422–25, for details.

32. Prosper Enfantin, *Colonisation de l'Algérie* (Paris: P. Bertrand, 1843), 63, 219.

33. Ibid., 70.

34. See Louis Moll, *Colonisation et agriculture de l'Algérie* (Paris: Librairie Agricole de la Maison Rustique, 1845), 108, and the discussion of this quote above.

35. For a typical example, see Marie Armand d'Avezac, "Esquisse générale de l'Afrique et Afrique ancienne," in *L'Afrique* (Paris: Firmin Didot Frères, 1844), 1–48.

36. Ibid., 34.

37. This is a significant departure from his earlier work, in which the period of Arab domination of North Africa is not portrayed as a destructive one, especially with regard to the environment. See for example Ernest Carette, *Études sur la Kabilie*, vol. 4 in the subseries Sciences historiques et

géographiques of the series Exploration scientifique de l'Algérie (Paris: Imprimerie Nationale, 1848), 451–53.

38. Ernest Carette, *Recherches sur l'origine et les migrations des principales tribus de l'Afrique septentrionale et particulièrement de l'Algérie,* vol. 3 in the subseries Sciences historiques et géographiques of the series Exploration scientifique de l'Algérie (Paris: Imprimerie Impériale, 1853), 397–98.

39. Ibid., 412.

40. Ibid., 413.

41. There are several translations of Idrisi's work, also known as the *Book of Roger,* and the translation and publication history of his writings, including the *Description of Africa,* is complex. A Latin edition can be dated to 1619 and a French translation to 1836. See Idrisi, *Description de l'Afrique et de l'Espagne,* trans. Reinhart P. A. Dozy and Michel Jean de Goeje (Leiden: E. J. Brill, 1968 [1866]), i–xxiii. Idrisi is also written as Edrisi or al-Idrissi. Idrisi (1099–1180) was most likely born in Northern Morocco. He was educated in Spain and subsequently moved to the royal court of King Roger of Sicily where he worked as a geographer. He completed the first edition of his book in 1154. Idrisi is cited very rarely in later iterations of the colonial declensionist environmental narrative.

42. Ibn Khaldoun, *Histoire des berbères et des dynasties musulmanes de l'Afrique septentrionale,* trans. William MacGuckin de Slane (Alger: Imprimerie du Gouvernement, 1852–56). De Slane (1801–78) was from Belfast, Ireland, and became a naturalized French citizen. His translation of this section of Ibn Khaldoun's work was extraordinarily influential and has been described as the "greatest textual event in the history of French Orientalism." See Abdelmajid Hannoum, "Translation and the Colonial Imaginary: Ibn Khaldûn Orientalist," *History and Theory* 42, no. 1 (February 2003): 62.

43. Ibn Khaldoun [Khaldûn] was born in Tunis to an aristocratic family, originally from Spain, in May of 1332. Highly educated, he studied briefly in Fez and then spent a few years in Spain. He subsequently returned to North Africa and later worked for the Moroccan Sultan ʿAbdul-ʿAziz pacifying "Arab tribes." He then lived several years during the 1370s in Algeria, where he wrote much of his book. Ibn Khaldoun later served the sultan in his native Tunis before making the pilgrimage to Mecca. He lived the last years of his life in Egypt, where he died in 1406. For more details, see Ibn Khaldoun, *The Muqaddimah: An Introduction to History,* trans. Franz Rosenthal, ed. N. J. Dawood (Princeton, N.J.: Princeton University Press, 1967), vii–ix. Ibn Khaldoun lived and wrote during a period of political upheaval and general decline in North Africa, one that followed several centuries of the Maghreb having been the

center of Medieval Arab civilization at its height. For an excellent discussion of the historical context, see Yves Lacoste, *Ibn Khaldun: The Birth of History and the Past of the Third World,* trans. David Macey (London: Verso, 1984).

44. Adolphes Noël Desvergers, *Histoire de l'Afrique sous la dynastie des aghlabites, et de la Sicile sous la domination musulmane* (Paris: F. Didot Frères, 1841). Carette cited Desvergers' translation in his 1853 volume on the principal tribes of Africa.

45. *Histoire des berbères et des dynasties musulmanes de l'Afrique septentrionale,* vol. 1, trans. Baron William MacGuckin de Slane, ed. Paul Casanova (Paris: Librairie Orientaliste Paul Geuthner, 1925), 34.

46. Ibid., 45.

47. Ibid., 165.

48. See Ibn Khaldoun, *Muqaddimah,* 91–118. As a member of the urban elite, it is not surprising that Ibn Khaldoun would have some very negative opinions about nomads. Urban biases against nomads are well documented in history and continue to exist to the present. Ibn Khaldoun's complicated life and wide variety of experiences in rural areas as well as urban spaces, though, likely tempered his views of nomads and probably helped to shape some of his favorable analyses of nomads and their ways of life.

49. Recent research has also shed light on the selective translation of such indigenous authors by colonial writers. Abdelmajid Hannoum, for example, shows that the language de Slane chose to use when he translated the *Histoire des berbères* essentially transformed the text in several ways, by imposing French colonial categories that were not justified by Ibn Khaldoun's original text. See Hannoum, "Translation."

50. See Lacoste, *Ibn Khaldun,* chap. 4. This topic was hotly debated by French scholars in the 1960s. See also Jean Poncet, "Le Mythe de la 'catastrophe' Hilalienne," *Annales: Économies, Sociétés, Civilisations* 22, no. 5 (1967): 1099–1120; and Roger Idris, "De la réalité de la catastrophe hilâlienne," *Annales: Économies, Sociétés, Civilisations* 23, no. 2 (1968): 390–96.

51. Lacoste, *Ibn Khaldun,* 65, quoting French historian Charles-André Julien.

52. Poncet, "Le Mythe," 1099, citing Roger Idris. Poncet argued in this and other articles that, rather than being a catastrophic invasion, the arrival of the Hillalians was more like "a nomad tide enlarging its waves gradually" over decades, and that it was not destructive. Instead, he shows that the Hillalians encountered a society in flux, beset by a series of political and economic crises due to other causes. The destruction of gardens and trees that Ibn Khaldoun attributed to the Hillalians, Poncet argued, was rather caused

by the troops of the local Emir during one of their sieges. Furthermore, Lacoste points out that, contrary to the often-cited "one million strong" nomad horde, the most reliable estimates point to only about fifty thousand nomads arriving in all of North Africa at this time. Lacoste, *Ibn Khaldun,* 71.

53. Émile F. Gautier, *Le Passé de l'Afrique du nord: Les Siècles obscurs* (Paris: Payot, 1942), 412. Gautier's book was especially important in spreading and popularizing the myth of the Arab invasion because it was widely read among French elites and the educated classes between the two world wars. Not only was it accepted by most intellectuals for about thirty years, but it also played a fundamental role in the attitudes of the French elites during the Algerian war for independence. I am grateful to Paul Claval for bringing the broader impact of Gautier's work to my attention.

54. Lacoste, *Ibn Khaldun,* 67.

55. Ibid., 75–76, 78. For further discussion of how this distinction between Arab and Berber was created and used, see Hannoum, "Translation and the Colonial Imaginary."

56. Lorcin, *Imperial Identities,* 2. Lorcin, though, disagrees with Lacoste's assertion that the dichotomy was used to "divide and rule" the Arabs and Berbers in Algeria.

57. Historian Edmund Burke has also written about the Kabyle myth in his frequently cited piece on the Moroccan Vulgate. See Edmund Burke, "The Image of the Moroccan State in French Ethnological Literature: A New Look at the Origin of Lyautey's Berber Policy," in *Arabs and Berbers: From Tribe to Nation in North Africa,* ed. Ernest Gellner and Charles Micaud (Lexington, Mass.: Lexington Books, 1972), 175–99. Ageron was probably one of the first to write of the Kabyle myth. See Charles-Robert Ageron, *Les Algériens musulmans et la France (1871–1919),* 2 vols. (Paris: Presses Universitaires de France, 1968).

58. Forester Paul Boudy claimed, as late as 1948, that the invading eleventh-century Arabs had "burned the forests that their innumerable herds then made disappear, leaving a bare space everywhere they passed." Paul Boudy, *Économie forestière nord-africaine: Milieu physique et milieu humain,* vol. 1 (Paris: Larose, 1948), 227.

59. This classification of Berbers as settled and Arabs as nomadic persisted during the colonial period despite the fact that some segments of the Arab population, particularly in the oases, for example, were settled, and some of the nomadic groups were Berber. Furthermore, although the Kabyles received the most attention as *the* Berber population in colonial Algeria, there were other Berber-speaking groups in Algeria, including the Tuareg, a nomadic

pastoral people. As a result, references to Arabs in colonial Algeria usually implied nomads or former nomads, especially in a rural context.

60. Fromentin heightens the sense of the pastoral by placing a flute in the hands of one of the shepherds in this lushly idyllic landscape setting. See Robert Canfritz, Lawrence Gowing, and David Rosand, *Places of Delight: The Pastoral Landscape* (Washington, D.C.: National Gallery of Art, 1988).

61. M. Oscar Mac Carthy, *Géographie physique, économique et politique de l'Algérie* (Alger: Dubos Frères, 1858), 106, see also 160.

62. When describing the nonforested areas, in fact, Mac Carthy praised the pastures, claiming that "there are few regions of the world as favored with natural fodder production." Ibid., 167.

63. Jules Duval, *L'Algérie: Tableau historique, descriptif et statistique* (Paris: L. Hachette, 1859), 26–27.

64. Ibid., 52.

65. Ibid., 7.

66. See Lorcin, *Imperial Identities,* 88–92, for a concise discussion of Urbain and his influence. See also Julien, *Histoire,* 256–58, 422–26.

67. He described at length, for example, the "fatal consequences" of their way of life: "[A]griculture struck by sterility, herds substituted for agriculture, no more tree plantations . . . the tent in place of a roof." Urbain published this book under a pseudonym. See Georges Voisin, *L'Algérie pour les Algériens* (Paris: Michel Lévy Frères, 1861), 34.

68. Ibid., 118.

69. Céleste Duval, *Moyens de rendre aux côtes nord d'Afrique tous les principes de fécondité dont elles sont susceptibles* (Paris: Napoléon Chaix, 1859), 21.

70. Ibid., 14.

71. The Forest Service had been decentralized and without a director since 1849, when it was divided into three parts corresponding to the three Algerian departments. Each of these had a different inspector and all were subordinated to the local administration operating in the civilian and military territories. See Boudy, *Économie forestière,* 1:232.

72. Auguste Mathieu, *Flore forestière: Description et histoire des végétaux ligneux qui croissent spontanément en France et des essences importantes de l'Algérie,* 2nd ed. (Nancy: Ancienne Maison Grimblot, 1860).

73. Ernest Cosson and Durieu de Maisonneuve, Ernest Cosson and M. Durieu de Maisonneuve, *Flore d'Algérie: Phanérogamie; Groupe des glumacées,* vol. 2 in the subseries Sciences naturelles: Botanique of the series Exploration scientifique de l'Algérie (Paris: Imprimerie Impériale, 1855–67), xlix–lv.

74. They explained, for instance, that the single tree that truly grows (naturally) in the high plateaus, *Pistacia atlantica,* is the "only [tree] that resists the violence of the winds and the variability of the temperature of these elevated plains." Ibid., lii.

75. Ibid., liii–liv.

76. Ibid., lv. This law they derived from the budding discipline of botanical geography, which is discussed further in chap. 4.

77. François Trottier, *Notes sur l'eucalyptus et subsidiairement sur la nécessité du reboisement de l'Algérie* (Alger: F. Paysant, 1867), 16.

78. Ibid., 10. He further explained that "fires and grazing are not the only causes of deforestation. . . . The Arab debarks the trees [for tannin] as they stand. . . . The tree dries out and, if the root does not die following this operation, it never grows to more than a shrub without value." Ibid., 11–12.

79. François Trottier, *Boisement et colonisation: Rôle de l'eucalyptus en Algérie au point de vue des besoins locaux, de l'exportation, et du développement de la population* (Alger: Association Ouvrière, 1876), 24. Trottier is here quoting an influential physician in Algeria, Dr. Bodichon, with whom he wholeheartedly agreed.

80. Trottier expected to earn 6,536 francs per hectare from his mature eucalyptus plantations. With the 84 hectares given to him by the government, he therefore hoped to earn 549,024 francs every decade (or about 55,000 francs per year). This was an extraordinary amount at the time given that a large and prosperous farm in the region was worth about 35,000 francs and rich men could retire well on 4,000 francs per year. See ibid., and Jean Bernis, *Quelques considérations sur la colonisation de l'Algérie* (Toulouse: Ph. Montaubin, 1866), 30.

81. Henri Verne, *La France en Algérie* (Paris: Charles Douniol, 1869), 20. Although he didn't specifically cite Ibn Khaldoun (or any source) for this description, he did cite Baron de Slane's translation of the *Histoire des berbères* a few pages later, and so was familiar with the work.

82. Ibid.

83. Ibid.

84. By introducing an insect which destroyed one of the most important forage plants for livestock, for example, the French pauperized and sedentarized most of the nomads of southern Madagascar. See Jeffrey Kaufmann, "The Cactus Was Our Kin: Pastoralism in the Spiny Desert of Southern Madagascar," in *Changing Nomads in a Changing World,* ed. Joseph Ginat and Anatoly Khazanov (Brighton: Sussex Academic, 1998), 124–42. See also Diana K. Davis, "Environmentalism as Social Control? An Exploration of the Transforma-

tion of Pastoral Nomadic Societies in French Colonial North Africa," *Arab World Geographer* 3, no. 3 (2000): 182–98.

85. For a good example of this forestcentric sentiment in the United States, see Shaul E. Cohen, "Promoting Eden: Tree Planting as the Environmental Panacea," *Ecumene* 6, no. 4 (1999): 424–46.

86. For an excellent discussion of the early roots of environmentalism, see Richard H. Grove, *Green Imperialism: Colonial Expansion, Tropical Island Edens, and the Origins of Environmentalism, 1600–1860* (Cambridge: Cambridge University Press, 1995). For changes in thinking about nature taking place in France, see Caroline Ford, "Nature, Culture and Conservation in France and Her Colonies 1840–1940," *Past and Present* 183, no. 1 (2004): 173–98.

87. George Perkins Marsh, *Man and Nature: Or, Physical Geography as Modified by Human Action,* ed. David Lowenthal (Cambridge, Mass.: Belknap Press, 1965), 114. Marsh, an American, had traveled widely throughout Europe and the Mediterranean (including Palestine and Egypt), and had served as minister to Turkey and ambassador to Italy. Ibid., xiii–xiv.

88. Louis Ferdinand Alfred Maury, *Histoire des grandes forêts de la Gaule et de l'ancienne France* (Paris: A. Leleux, 1850); and Antoine César Becquerel, *Des Climats et de l'influence qu'exercent les sols boisés et non boisés* (Paris: Firmin Didot Frères, 1853).

89. Marsh, *Man and Nature,* 117. The idea that the Sahara Desert should be forested is still influential in certain policy circles. In 1982, for example, Mauritania, Morocco, Algeria, Tunisia, and Libya agreed to cooperate on implementing "a transcontinental forest extending through the territory" of these five countries. See Florian Plit, "La Dégradation de la végétation, l'érosion et la lutte pour protéger le milieu naturel en Algérie et au Maroc," *Méditerranée* 49, no. 3 (1983): 87. This idea has recently been revived, especially in Algeria, where the government envisions a trans-Saharan "green belt from the Atlantic to the Red Sea." See République Algérienne Democratique et Populaire, Direction des Forêts, *Rapport national de l'Algérie sur la mise en oeuvre de la convention de lutte contre la désertification* (Alger: Ministère de l'Agriculture et du Developpement Rural, Direction des Forêts, 2004), 6.

90. *Man and Nature* was published simultaneously in the United States and England and was translated into Italian within five years of its original publication. Marsh, *Man and Nature,* xxvii.

91. Ibid., xxii. See also Paul Claval, "French Perceptions of Nature over the Last Centuries" (unpublished manuscript, 2005).

92. Trottier, *Notes sur l'eucalyptus,* 14, see also 17.

93. For a masterful overview of changing ideas of nature over the last two millennia, see Clarence Glacken, *Traces on the Rhodian Shore* (Berkeley: University of California Press, 1967). See also Robert P. Harrison, *Forests: The Shadow of Civilization* (Chicago: University of Chicago Press, 1992), which provides a partial outline of changing ideas about forests over time.

94. Glacken, *Traces*, 491–94.

95. John Evelyn, *Silva: Or, A Discourse of Forest-Trees* (York: A. Ward, 1776).

96. For early conservation, see Glacken, *Traces*, 322–30; and John F. Freeman, "Forest Conservancy in the Alps of Dauphiné, 1287–1870," *Forest and Conservation History* 38 (October 1994): 171–80.

97. Richard H. Grove, *Ecology, Climate, and Empire: Colonialism and Global Environmental History, 1400–1940* (Cambridge: White Horse, 1997), 196. Similar edicts were passed in parts of France, as early as the ninth century, to stop the extension of forests. See Hippolyte Doumenjou, *Études sur la révision du code forestier: Les Reboisements en France et en Algérie* (Paris: J. Baudry, 1883), 300.

98. Glacken, *Traces*, 336.

99. Ibid., 130, 121. To understand what Plato may have meant and how various interpretations of his work have been exaggerated, see A. T. Grove and Oliver Rackham, *The Nature of Mediterranean Europe: An Ecological History* (New Haven, Conn.: Yale University Press, 2001), esp. 288.

100. Stephen J. Pyne, *Vestal Fire: An Environmental History, Told through Fire, of Europe and Europe's Encounter with the World* (Seattle: University of Washington Press, 1997), 112.

101. Glacken, *Traces*, 345.

102. Pyne, *Vestal*, 107. In fact, France pioneered some of the earliest forest protection and regulatory policies and practices in the Western world, dating back to the seventh century. See Georges Plaisance, *La Forêt française* (Paris: Édition Denoël, 1979).

103. See Glacken, *Traces*; and Pyne, *Vestal Fire*.

104. For an enlightening discussion of this idea in the French context, see Emma C. Spary, *Utopia's Garden: French Natural History from Old Regime to Revolution* (Chicago: University of Chicago Press, 2000), esp. 102–17.

105. Comte, quoted in John Noyes, "Nomadic Fantasies: Producing Landscapes of Mobility in German Southwest Africa," *Ecumene* 7, no. 1 (2000): 50.

106. Marquis de Condorcet, *Esquisse d'un tableau historique des progrès de l'esprit humain* (Paris: Agasse, 1794). Condorcet also posited a hypothetical future tenth stage that would be a utopian era of equality, peace, and plenty. It should be noted that in the fourteenth century Ibn Khaldoun also wrote

on the passage of humankind from a nomadic stage to a sedentary urban stage. He thought, however, that the sedentary urban stage would ultimately decay and fall, to be replaced by the nomadic stage once again. Many have interpreted this to be a cyclical process in which the nomadic stage is the stage of renewal from which the progression is born, and therefore to be perceived as a "good" and not a "bad" stage. See Ibn Khaldoun, *Muqaddimah*, esp. 91–95.

107. For a good discussion of these ideas, see Noyes, "Nomadic Fantasies."

108. Spencer, quoted in Jean Comaroff and John Comaroff, *Of Revelation and Revolution: The Dialectics of Modernity in a South African Frontier* (Chicago: University of Chicago Press, 1997), 2:213.

109. Noyes, "Nomadic Fantasies," 51.

110. Verne, *La France*, 36.

111. Grove, *Green Imperialism*.

112. See Jerome H. Buckley, *The Triumph of Time: A Study of the Victorian Concepts of Time, History, Progress, and Decadence* (Cambridge, Mass.: Belknap Press, 1966); and David R. Stoddart, *On Geography and Its History* (Oxford: Basil Blackwell, 1986).

113. Richard H. Grove, "Scottish Missionaries, Evangelical Discourses and the Origins of Conservation Thinking in Southern Africa 1820–1900," *Journal of Southern African Studies* 15, no. 2 (1989): 180–81. See also Georgina H. Endfield and David J. Nash, "Drought, Desiccation and Discourse: Missionary Correspondence and Nineteenth-Century Climate Change in Central Southern Africa," *Geographical Journal* 168, no. 1 (2002): 33–47.

114. Quoted in Paulin Trolard, *Bréviaire du ligeur* (Alger: Casabianca, 1893), 4.

115. André Nouschi, *Enquête sur le niveau de vie des populations rurales constantinoises de la conquête jusqu'en 1919* (Paris: Presses Universitaires de France, 1961), 82.

116. Auguste Warnier, *L'Algérie devant l'empereur, pour faire suite à l'Algérie devant le sénat, et à l'Algérie devant l'opinion publique* (Paris: Challamel Ainé, 1865), 28.

117. J. Reynard, *Restauration des forêts et des pâturages du sud de l'Algérie (province d'Alger)* (Alger: Adolphe Jourdan, 1880), 15.

118. Ligue du Reboisement, *La Forêt: Conseils aux indigènes* (Alger: Association Ouvrière P. Fontana, 1883), 2.

119. See, for instance, Michael J. Heffernan, "The Desert in French Orientalist Painting during the Nineteenth Century," *Landscape Research* 16, no. 2 (1991): 37–42; and Jean-Claude Vatin, "Désert construit et inventé, Sahara perdu ou retrouvé: Le Jeu des imaginaires," in *Le Maghreb dans l'imaginaire*

français: La Colonie, le désert, l'exil, ed. Jean-Robert Henry (Aix-en-Provence: La Calade, 1985), 107–31.

120. In 1859 the French author Théophile Gautier remarked that "today the Sahara is dotted with as many landscapists' parasols as the Forests of Fontainebleau in days gone by." Roger Benjamin, ed. *Orientalism: Delacroix to Klee* (Sydney: The Art Gallery of New South Wales, 1997), 14. For an insightful discussion of Gautier's relationship to orientalism, see James E. Housefield, "Orientalism as Irony in Gérard de Nerval's *Voyage en Orient,*" *Journal of North African Studies* 5, no. 4 (2000): 10–24.

121. This painting garnered the praise of Gautier and of an inspector of the fine arts office when it was displayed, but it also drew some criticism. For more information, see Institut du Monde Arabe, *De Delacroix à Renoir: L'Algérie des peintres* (Paris: Institut du Monde Arabe/Hazan, 2003), 172.

122. Belly, quoted in Christine Peltre, *Orientalism in Art,* trans. John Goodman (New York: Abbeville, 1998), 162.

123. Heffernan, "Desert," 40–41.

124. The original painting shows a range of colors, shadow, and light that are more consistent with sunset than sunrise. Such details are generally lost in reproductions of Guillaumet's canvas. Special thanks to art historian James Housefield, with whom I developed this interpretation. Roger Benjamin's outstanding recent book corroborates this argument. Roger Benjamin, *Orientalist Aesthetics: Art, Colonialism, and French North Africa, 1880–1930* (Berkeley: University of California Press, 2003), 53.

125. Anne-Marie Delage, "Guillaumet, Gustave," in *Grove Art Online,* http://www.groveart.com. Later in the colonial period, the government of Jules Ferry in fact actively funded orientalist artists with traveling scholarships and money to support "Orientalist painting displays in the colonial pavilions at the Universal Exhibitions of 1889 and 1900." The Ferry government, especially influential senators like Eugène Étienne who strongly supported French rule in Algeria, "understood the potential of Orientalist painting and sculpture for popularizing the colonies." Benjamin, ed., *Orientalism,* 20.

126. Two versions exist. The painting usually considered the earlier is self-consciously inspired by Théodore Géricault's 1819 *Raft of the Medusa.* Heffernan, "Desert," 39–40.

127. Eugène Fromentin, *Sahara et Sahel. I.—Une été dans le Sahara. II.— Une année dans le Sahel,* illustrated ed. (Paris: E. Plon, 1879), 188.

128. See ibid., 186.

129. Heffernan, "Desert," 40.

130. Fromentin, quoted in Bernard and Lacroix, *L'Évolution,* 44.

131. *Une été dans le Sahara,* originally published in 1857 (unillustrated), was written first and was very popular. The first time it was paired with *Une année dans le Sahel* appears to have been in the 1879 illustrated edition. An English translation of *Une année dans le Sahel* appeared in 1999 and was reissued as a trade paperback in 2004, attesting to the continued popularity of Fromentin's desert writings. See Eugène Fromentin, *Between Sea and Sahara: An Algerian Journal,* trans. Blake Robinson (Athens: Ohio University Press, 1999).

132. Both paintings also stand out as unusually bleak and despairing works in the oeuvres of both Guillaumet and Fromentin. In fact, they are remarkable within the larger scope of orientalist desert painting for their grim and gloomy tone. They should not be discounted, however, because of this. Rather, they should be considered within the context of their time and especially of the dominant colonial environmental narrative of the period.

133. Buckley, *Triumph,* 28.

134. Stoddart, *On Geography,* 160–67. This expanded notion of history took even longer to spread to those working in the North African territories, perhaps because it was not often useful in the propaganda designed to promote the colonial venture.

135. See Grove, *Green Imperialism.*

136. Trottier, *Boisement et colonisation,* 32.

137. See John Croumbie Brown, *French Forest Ordinance of 1669; with Historical Sketch of Previous Treatment of Forests in France* (Edinburgh: Oliver and Boyd, 1883); and Alexandre Surell, *Étude sur les torrents des hautes-Alpes, avec une suite par Ernest Cézanne, tome second,* 2nd ed. (Paris: Dunod, 1872), preface to vol. 2. Particularly important forest protection edicts were proclaimed in 1215, 1219, 1291, 1346, 1376, 1537, and 1597, among other years.

138. For details on the 1669 ordinance and its contents, see Gustave Huffel, *Économie forestière, tome premier, deuxième partie: Propriété et législation forestières, politique forestière, la France forestière, statistiques.* 2nd ed. (Paris: Librairie Agricole de la Maison Rustique, 1920), 290–94.

139. See Glacken, *Traces,* 491–94; and Brown, *French Forest Ordinance.*

140. See Whited, *Forests and Peasant,* 23.

141. These officers, assigned to different regions, were responsible for visiting their forests, making sure the ordinance was enforced, and punishing offenders.

142. A few years later, France was divided into 28 forest districts. For details on this period, see Huffel, *Économie forestière,* 297–310.

143. In the nineteenth century, the Forestry School at Nancy, one of the first in Europe, was well respected internationally and trained many foreign

foresters, including 81 British foresters between 1867 and 1893. See Charles Guyot, *L'Enseignement forestier en France: L'École de Nancy* (Nancy: Crépin-Leblond, 1898), 192–93. Many of these British foresters worked in India and other British colonial possessions. In fact, the influence of the British foresters trained in French methods at Nancy was so strong that Rudyard Kipling parodied the Indian Forest Service by writing that "the reboisement of all India is in its hands. . . . Its servants wrestle with . . . coarse grass and unhappy pine after the rules of Nancy." Rudyard Kipling, "In the Rukh," in *Many Inventions* (New York: D. Appleton, 1899), 222–23.

144. Whited, *Forests and Peasant*, 34.

145. Ibid., 35. Whited provides an excellent discussion and overview of the 1827 Forest Code and reaction to its implementation; see 34–46. See also Plaisance, *La Forêt*, 201–2.

146. Whited, *Forests and Peasant*, 35.

147. The Forest Administration was composed of a hierarchy of five senior administrators, thirty conservators, two hundred inspectors, three hundred subinspectors, five hundred general forest guards, and eight thousand local forest guards. In addition to the National Forestry School at Nancy, which trained most of the intermediate-level general forest guards and subinspectors, three schools for forest guards were founded in the 1860s. The lowest positions, those of local forest guards, were often filled by men from the military rather than the forestry schools. See ibid., 30–34, for details.

148. For an interesting study of this event, see Peter Sahlins, *Forest Rites: The War of the Demoiselles in Nineteenth-Century France* (Cambridge, Mass.: Harvard University Press, 1994).

149. See Pyne, *Vestal Fire*, 112–15.

150. Whited, *Forests and Peasant*, 56–63, provides a good discussion of Surell and his impact in France. Surell's 1841 text enjoyed a second edition that was expanded to two volumes, the second of which was written by Ernest Cézanne. See Surell, *Étude*, vol. 1.

151. Jean Antoine Fabre, *Essai sur la théorie des torrens et des rivières* (Paris: Bidault, 1797).

152. See Glacken, *Traces*, 698–702, for an excellent overview of Fabre's work.

153. Grove, *Green Imperialism*, 259.

154. Whited, *Forests and Peasant*, 57–58.

155. The text of this law and a discussion of it may be found in Surell, *Étude*, 2:182–213 and 373–77. The law made it necessary that new courses be added to the curriculum at the Forestry School at Nancy to teach reforestation. See Guyot, *L'Enseignement forestier*, 211.

156. Plaisance, *La Forêt*, 202.

157. In fact, the idea that human action could change the climate is quite old and can be traced at least to the Greeks and the writings of Theophrastus. See Glacken, *Traces*, 129–30 and 270.

158. For a detailed discussion of the English and French scientific developments related to desiccation theory, see Grove, *Green Imperialism*, particularly 158–67.

159. Buffon, *Histoire naturelle*, as paraphrased in Glacken, *Traces*, 669–70.

160. See Grove, *Green Imperialism*.

161. Marsh, *Man and Nature*, 158.

162. Jules Clavé, *Études sur l'économie forestière* (Paris: Guillaumin, 1862), 46, as quoted in Marsh, *Man and Nature*, 160. Clavé provides another typical description of desiccation in his discussion of the French forest of Fontainebleau. See Jules Clavé, "Études forestières: La Forêt de Fontainebleau," *Revue des Deux Mondes* 45 (May-June 1863): 162–63.

163. J.-A.-N. Périer, *De l'hygiène en Algérie*, vol. 1 in the subseries Sciences médicales of the series Exploration scientifique de l'Algérie (Paris: Imprimerie Royale, 1847), 14, 21.

164. Reynard, *Restauration des Forêts*, 6.

165. See J. E. Planchon, "L'Eucalyptus globulus au point de vue botanique, économique et médical," *Revue des Deux Mondes* 7, no. 1 (January 1875): 162–63.

166. See Grove, *Green Imperialism*, for several detailed examples of this activity.

167. The relationship between deforestation and erosion would not significantly impact Algerian forestry until the early twentieth century. See, for example, Charles Rabot, "Le Déboisement et l'aggravation de la torrentialité en Algérie," *La Géographie, Bulletin de la Société de Géographie* 11, no. 15 (July 1905): 315–17.

168. François Combes, quoted in Whited, *Forests and Peasant*, 63.

169. Ibid., 64.

170. Vasant Saberwal, "Science and the Desiccationist Discourse of the 20th Century," *Environment and History* 4, no. 3 (1998): 321, see also 316.

171. Stephen E. Darby, "Effects of Riparian Vegetation on Flow Resistance and Flood Potential," *Journal of Hydraulic Engineering* 125, no. 5 (1999): 443–54.

172. Ibid., 450–53.

173. For a typical example of nineteenth-century arguments for desiccation theory, based on nonexperimental (anecdotal) evidence, see Marsh, *Man and Nature*, 171–82. Much of the faith in the idea of deforestation causing the

reduction of streams and lake levels came from the publication of Alexander von Humboldt's essays on his travels in South America at the turn of the nineteenth century (1799–1804). In his *Personal Narrative of Travels to the Equinoctial Regions of America*, Humboldt described the lake of Valencia in Venezuela as having dried up because many of the trees had been cleared in the valley in which it was located. In fact, however, the streams feeding Lake Valencia were at that time diverted for irrigation by the people living in the valley. When the French agricultural chemist Joseph Boussingault visited the same lake twenty-two years later, he found that "the waters of the lake were no longer retiring, but, on the contrary, were sensibly rising. . . . At the time of the growing prosperity of the valley . . . the principal affluents of the lake [had been] diverted, to serve for irrigation. . . . At the period of my visit, their waters, no longer employed, flowed freely." Boussingault, quoted in Marsh, *Man and Nature*, 176–77. Despite his observations, Boussingault did believe in the theory that deforestation caused desiccation. See Jean Baptiste Boussingault, *Économie rurale considérée dans ses rapports avec la chimie, la physique et la météorologie*, 2nd ed. (Paris: Béchet Jeune, 1851), 2:730–59.

174. Saberwal, "Science," 316–17.

175. Ibid., 317.

176. J. M. Bosch and J. D. Hewlett, "A Review of Catchment Experiments to Determine the Effects of Vegetation Changes on Water Yield and Evapotranspiration," *Journal of Hydrology* 55, no. 1–4 (1982): 3–23.

177. It is this property of eucalyptus of using large amounts of water that made it so useful to the French in helping to dry out marshy areas in Algeria, particularly in the Mitidja. See Julien Franc, *La Colonisation de la Mitidja* (Paris: Honoré Champion, 1928). The biophysical impacts of eucalyptus are controversial but many environmentalists claim that it depletes soil moisture, sterilizes the ground around plantations, and increases erosion on the bare ground below the trees, where nothing else can grow. For a useful overview of the debate, see Robin W. Doughty, *The Eucalyptus: A Natural and Commercial History of the Gum Tree* (Baltimore: Johns Hopkins University Press, 2000), 144–58.

178. Saberwal, "Science," 316.

179. See Grove, *Green Imperialism*, 209.

180. See René Louiche Desfontaines, "Fragmens d'un voyage dans les régences de Tunis et d'Alger, fait de 1785 à 1786," in *Voyages dans les régences de Tunis et d'Alger*, ed. M. Dureau de la Malle (Paris: Gide, 1838), 2:261. Desfontaines also apparently did not believe that the Sahara Desert was created by wanton deforestation.

181. See, for example, Endfield and Nash, "Drought."

182. For rainfall data, see Sharon E. Nicholson, "The Historical Climatology of Africa," in *Climate and History*, ed. T. Wigley, M. Ingram, and G. Farmer (Cambridge: Cambridge University Press, 1981), 249–70; and Sharon E. Nicholson, "Recent Rainfall Fluctuations in Africa and their Relationship to Conditions over the Continent," *Holocene* 4, no. 2 (1994): 121–31. Clavé was likely citing the report of the duc de Raguse on his voyage in the 1830s to Egypt. See duc de Raguse, *Voyage du maréchal duc de Raguse* (Paris: Ladvocat, 1837).

183. Prosper Demontzey, the "father" of reforestation in France, began his reforestation career as a guard in Algeria in 1853, shortly after graduating from the forestry school at Nancy. He directed a well-known reforestation project at Orléansville. He worked in Algeria for nearly a decade and returned to France in 1862 to head the reforestation service in the Alpes-Maritimes. See Association des Ingénieurs du Génie Rural, des Eaux et des Forêts [hereafter, AIGREF], *Des Officiers royaux aux ingénieurs d'état dans la France rurale (1219–1965)* (Paris: Éditions Technique et Documentation, 2001), 104–5, 177; and Boudy, *Économie forestière*, 1:233–34, 477. The influence of his experiences in Algeria on Demontzey's subsequent work on reforestation in France remains, as far as I can find, to be explored.

184. Victor Renou, considered the founder of the Algerian forestry service, graduated from the National Forestry School at Nancy. He died prematurely in 1844, only five years after the service was created, after being thrown from his horse while returning from a forest inspection near Bône. See Boudy, *Économie forestière*, 1:230, 232.

185. Ageron, *Les Algériens*, 1:107. See also Angel-Paul Carayol, *La Législation forestière de l'Algérie* (Paris: Arthur Rousseau, 1906).

186. AIGREF, *Officiers royaux*, 177.

187. Whited, *Forests and Peasant*, 31.

188. For details see AIGREF, *Officiers royaux*, 177, and Boudy, *Économie forestière*, 1:232–33.

189. Ageron, *Les Algériens*, 1:107.

190. See Boudy, *Économie forestière*, 1:233, 235.

191. Governor-General Bugeaud outlined the use of collective punishment to punish "unruly tribes" in his memorandum of 2 January 1844. In a separate memorandum from 1840 he authorized "sequestration," or state seizure of the property of "rebellious tribes." This became a government regulation in 1845 and remained in effect into the twentieth century. See Carayol, *La Législation*, 38–39.

192. V. Boutilly, *Recueil de la législation forestière algérienne, lois, décrets et règlements divers* (Paris: Berger-Levrault, 1905), 281. This law had a precedent in a French law of 1830 that allowed the state to "expropriate private property needed to serve public utility." John E. Rodes, *A Short History of the Western World* (New York: Charles Scribner's Sons, 1970), 374.

193. AIGREF, *Officiers royaux*, 177.

194. Ageron, *Les Algériens*, 1:109–10. This amounted to the government practically giving away large areas of state cork forest to private owners and thus giving up the long-term rents the concessions had generated. These concessions were essentially long-term leases for which the concessionaires paid the Algerian government a fee. In the late 1840s, concessionaires paid on a sliding scale: for the first cork harvest they paid 10 percent of the market value, for the second harvest 15 percent, and for the third and fourth harvests, 30 percent. See Eugène Battistini, *Les Forêts de chêne-liège de l'Algérie* (Alger: Victor Heintz, 1937), 30. The leases began, for a short period, as sixteen-year leases, but over the course of the next few decades they became longer, first forty and then ninety years by 1862. See Henri Marc, *Notes sur les forêts de l'Algérie* (Alger: Adolphe Jourdan, 1916), 57; and A. D. Combe, *Les Forêts de l'Algérie* (Alger: Giralt, 1889), 28. Cork, the bark of the cork oak tree, is harvested every eight years or so and regrows between harvests if harvested correctly.

195. It is important to note that most of the forest concessions at this time were for cork oak. This species is very well adapted to fire because the bark (cork) is fire resistant. When these trees are harvested for their bark, however, they become quite susceptible to fire. Thus the harvesting of cork in the concession forests made them more susceptible to fire and helped to create the increasingly common conflagrations.

196. Ageron, *Les Algériens*, 1:111.

197. For details, see ibid., 1:112–13; Boudy, *Économie forestière*, 1:246–47; and Carayol, *La Législation*, 32–45.

198. Boudy, *Économie forestière*, 1:247, AIGREF, *Officiers royaux*, 178. Some in the Algerian government, however, worried that placing the Forest Service under metropolitan control would give it too much autonomy and power and thereby make certain aspects of government more difficult. See Ageron, *Les Algériens*, 1:116.

199. See Ageron, *Les Algériens*, 1:124; and Carayol, *La Législation*, 45–58.

200. Carayol argued that one of the goals of this 1885 forest law was to apply some of the 1882 French law on the restoration of mountain lands to Algeria, although it was not a perfect fit. See *La Législation*, 55.

201. AIGREF, *Officiers royaux*, 178.

202. Ibid.

203. Furthermore, article 71 forbade sheep, goats, and camels from grazing in state or communal forests (except by special dispensation from the governor-general), but allowed pigs, revenue earners for the French but anathema to local Muslims, to graze. Theodore Woolsey, *French Forests and Forestry: Tunisia, Algeria, Corsica* (New York: John Wiley, 1917), 178. Woolsey provides a useful English translation of the 1903 Algerian Forest Code; see 161–208. See also Carayol, *La Législation,* for more details.

204. Ageron, *Les Algériens,* 2:776.

205. See Boudy, *Économie forestière,* 1:477–78; and Woolsey, *French Forests,* 179. Despite all the concern over reforestation, only 5,400 hectares had been "reforested" by 1910.

206. Carayol, *La Législation,* 89.

207. Ibid., 263–64. The 1881 figure comes from Ageron, *Les Algériens,* 1:126. See also Combe, *Les Forêts,* 49.

208. Combe, *Les Forêts,* 9–10.

209. Ageron, *Les Algériens,* 1:106.

210. Ibid.

211. Ibid. Boudy estimated that at least one million hectares were "deforested" (he counts the loss of brush as well as trees) from 1870 to 1940 from both legal and illegal activities. Boudy, *Économie forestière,* 1:558–59.

212. This estimate of the extent of cork oak forest is from Battistini, *Les Forêts,* 20.

213. Ibid., 16.

214. Ibid., 26.

215. Marc, *Notes sur les forêts,* 57.

216. Boudy, *Économie forestière,* 1:236.

217. Ageron, *Les Algériens,* 1:109–10.

218. Julien, *Histoire,* 406–9 and 436–38, provides appalling details of the exploitative acts of some of these companies.

219. Ibid.

220. See ALG, GGA, 3L/32, CAOM, for information on the activities of the Société Générale Algérienne from the 1860s to the 1880s.

221. Mahfoud Bennoune, *The Making of Contemporary Algeria, 1830–1987: Colonial Upheavals and Post-independence Development* (Cambridge: Cambridge University Press, 1988), 48.

222. Ibid.

223. Ibid. See also table 2.4, 50.

224. Julien, *Histoire,* 379–81.

225. Ibid., 404–6. Historians have argued that, in effect, the French created "collective property" as a way to justify "Muslim tribal" lands as having no owner, only use rights, and therefore as in need of being titled. This was sleight of hand on the part of the colonial government, since private property did exist in Muslim Algeria. See ibid.; and Ageron, *Modern.*

226. Lapasset, quoted in Julien, *Histoire,* 405.

227. Ibid., 406.

228. Ibid., 422. Many of the colonists were furious that cantonment was made illegal, claiming that it would make colonization in Algeria impossible. See Auguste Warnier, *L'Algérie devant le sénat* (Paris: Dubuisson, 1863), 166.

229. Bennoune, *Making,* 45.

230. Ibid.

231. For details on this process see Julien, *Histoire,* 426–27; and Bennoune, *Making,* 44–45.

232. General Allard, quoted in Bennoune, *Making,* 44.

233. Anon., *Statistique et documents relatifs au Sénatus-Consulte sur la propriété arabe* (Paris: Imprimerie Impériale, 1863), 158.

234. Captain Vaissière, quoted in Bennoune, *Making,* 45.

235. See Louis-François Victor Tassy, *Service forestier de l'Algérie, rapport adressé à m. le gouverneur de l'Algérie* (Paris: A. Hennuyer, 1872), 5.

236. Wolfgang Trautmann, "The Nomads of Algeria under French Rule: A Study of Social and Economic Change," *Journal of Historical Geography* 15, no. 2 (1989): 126–38. See also M'hammed Boukhobza, *Monde rural: Contraintes et mutations* (Alger: Office des Publications Universitaires, 1992).

237. See Ageron, *Les Algériens,* 1:103–28; and Nouschi, *Enquête.*

238. Ageron, *Modern,* 44; and Lacoste, Nouschi, and Prenant, *L'Algérie,* 372–74.

Chapter 4

1. Charles-Robert Ageron, *Modern Algeria: A History from 1830 to the Present,* trans. Michael Brett (London: Hearst, 1991), 53.

2. For details of this period, see Roger Price, *A Concise History of France* (Cambridge: Cambridge University Press, 1993), 191–208. This "civil war" in France is considered the last of the great nineteenth-century revolutions.

3. The introduction of secular education was partly motivated, though, by the government's quest to diminish the power of the Catholic Church. Divorce was also legalized and religious education in primary schools was forbidden. Ibid.

4. Unions were given legal status in 1884 and the General Confederation of Labor was formed in 1895. Clashes between labor and the military punctuated the Third Republic. John E. Rodes, *A Short History of the Western World* (New York: Charles Scribner's Sons, 1970), 418.

5. Price, *Concise History,* 194.

6. Ferry, quoted in Robert Aldrich, *Greater France: A History of French Overseas Expansion* (London: Macmillan, 1996), 98.

7. Some French historians have pointed out, however, that in reality colonial possessions on the whole did not furnish significant markets for France. In 1913, for example, only about 13 percent of French exports were purchased by its colonies. Ibid., 196.

8. Aldrich, *Greater France,* provides a detailed overview of French colonial expansion.

9. Ibid., 101–2.

10. Ageron, *Modern,* 63.

11. For details of the Algerian Insurrection and its brutal handling by the French, see Charles-Robert Ageron, *Les Algériens musulmans et la France (1871–1919),* 2 vols. (Paris: Presses Universitaires de France, 1968); and Yves Lacoste, André Nouschi, and André Prenant, *L'Algérie: Passé et présent* (Paris: Éditions Sociales, 1960).

12. Ageron, *Modern,* 53.

13. John Ruedy, *Modern Algeria: The Origins and Development of a Nation* (Bloomington: Indiana University Press, 1992), 93.

14. Ageron, *Modern,* 54.

15. See Ageron, *Les Algériens,* for details on the changes forced on the indigenous Algerians during this period.

16. Ageron, *Modern,* 54.

17. Ruedy, *Modern,* 91.

18. Ageron, *Modern,* 56.

19. Mahfoud Bennoune, *The Making of Contemporary Algeria, 1830–1987: Colonial Upheavals and Post-independence Development* (Cambridge: Cambridge University Press, 1988), 48–50.

20. For details on the Warnier Law and its effects, see Ageron, *Les Algériens,* and Bennoune, *Making.*

21. Ageron, *Les Algériens,* 1:97. The Warnier Law ceased to be applied in 1892 and was abrogated by 1897.

22. Bennoune, *Making,* 47.

23. Ageron, *Les Algériens,* 2:741.

24. Ageron, *Modern,* 60.

25. Ibid., 62, 84. Wine remained one of the most important components of the Algerian economy production until the 1960s, when petroleum production began to generate large profits.

26. Ibid., 56.

27. See Ruedy, *Modern*, 86–87, and Ageron, *Modern*, 63–64.

28. See Ageron, *Modern*, 75.

29. Ibid., 60.

30. See Ruedy, *Modern*, 95.

31. Some sources estimate that less than 15 percent of Algerian land was owned by Algerians by the early twentieth century. Michael Heffernan and Keith Sutton, "The Landscape of Colonialism: The Impact of French Colonial Rule on the Algerian Rural Settlement Pattern, 1830–1987," in *Colonialism and Development in the Contemporary World*, ed. Christopher Dixon and Michael Heffernan (London: Mansell, 1991), 133.

32. Ruedy, *Modern*, 98.

33. Ibid.

34. For biographical information, see Patricia M. Lorcin, *Imperial Identities: Stereotyping, Prejudice and Race in Colonial Algeria* (London: I. B. Tauris, 1995), 129–30, 309.

35. It is little known that Warnier and Duval also wrote a monograph on the forests of Algeria that was never published. See Ernest Cosson, *Compendium florae atlanticae ou flores des états barbaresques, Algérie, Tunisie, Maroc*, vol. 1 (Paris: Imprimerie Nationale, 1881), 100.

36. Ruedy, *Modern*, 81.

37. To be sure, parts of the declensionist environmental narrative had served to help justify earlier laws that dealt with property, such as the 1851 land law and the 1863 Sénatus-consulte.

38. See, for example, the discussion in the last chapter of Henri Verne, *La France en Algérie* (Paris: Charles Douniol, 1869).

39. Auguste Warnier, *L'Algérie devant l'empereur, pour faire suite à l'Algérie devant le sénat, et à l'Algérie devant l'opinion publique* (Paris: Challamel Ainé, 1865), 5.

40. Ibid., 10.

41. Ibid., 28 (emphasis in original).

42. Ibid.

43. Ibid., 6.

44. Ibid.

45. Ibid., 9.

46. Ibid., 27–28.

47. For examples, see Auguste Warnier, *L'Algérie devant le sénat* (Paris: Dubuisson, 1863); and Jules Duval and Auguste Warnier, *Un Programme de politique algérienne* (Paris: Challamel Ainé, 1868). Interestingly, Warnier did not invoke the declensionist narrative in one of his earliest publications. See Ernest Carette and Auguste Warnier, *Description et division de l'Algérie* (Paris: L. Hachette, 1847).

48. See Ageron, *Les Algériens*, 1:78–84, for details on these debates and revisions.

49. Ibid., 1:79.

50. Quoted ibid., 1:78.

51. Ibid., 1:87–88.

52. Ibid., 1:93.

53. Ibid., 2:742.

54. Ibid., 2:741.

55. Bennoune, *Making*, 63–64.

56. Ageron, *Modern*, 60.

57. Bennoune, *Making*, 65.

58. Ageron, *Les Algériens*, 2:807, 813.

59. Ibid., 2:811.

60. The settlers were vocal and persistent in their efforts to acquire more and more land. See the pleading of one such group (whose motto was "By the French for the French") to the governor-general in J. Jourde, *La Ligue pour la défense de la colonisation sur les hauts-plateaux à monsieur Laferrière, gouverneur général de l'Algérie* (Saida: Émile Pascal, 1898).

61. Léon Lehureaux, *Le Nomadisme et la colonisation dans les hauts plateaux de l'Algérie* (Paris: Comité de l'Afrique Française, 1931), chaps. 2 and 3.

62. M'hammed Boukhobza, *Monde rural: Contraintes et mutations* (Alger: Office des Publications Universitaires, 1992), 26. Only about 5 percent remained nomads at the time of independence.

63. Wolfgang Trautmann, "The Nomads of Algeria under French Rule: A Study of Social and Economic Change," *Journal of Historical Geography* 15, no. 2 (1989): 126–38.

64. Ibid., 130. As a result of this overexploitation, special governmental decrees based on the advice of the Forest Service began to regulate the harvesting and sale of this plant in the High Plateaus in 1888. There had also been some provisions regulating alfa in the 1885 forest law. See Louis Trabut, *L'Halfa* (Alger: Giralt, 1889).

65. Augustin Bernard and Napoléon Lacroix, *L'Évolution du nomadisme en Algérie* (Alger: Adolphe Jourdan, 1906).

66. Ibid., 24

67. Ibid., 26.

68. Ibid., 44, 47.

69. Ibid., 5.

70. Ibid., 7.

71. Ibid., 20–21.

72. Ibid., 22.

73. Ibid., 43. For this information they quote Paulin Trolard's 1904 report to the special forestry commission. The significance of this is explained in the following section on the Ligue du Reboisement, a powerful group headed by Trolard that successfully lobbied for significant forestry reforms.

74. See Gouvernement Général de l'Algérie, *Création d'un service pastoral* (Alger-Mustapha: Giralt, 1896). Much of colonial "range management" in Algeria and later Tunisia and Morocco was based in forestry and veterinary medicine, as is discussed in the next chapter.

75. Bernard, *L'Évolution*, 302.

76. Ibid.

77. J. E. Planchon, "L'Eucalyptus globulus au point de vue botanique, économique et médical," *Revue des Deux Mondes* 7, no. 1 (January 1875): 149–74.

78. Ibid., 161. Ramel also enthusiastically spread seeds in many other parts of the Mediterranean. Due in large part to his enthusiasm and activity promoting the trees, France is considered one of the key points of distribution and spread of eucalyptus in the nineteenth century. Eucalyptus did very well in many parts of southern France and in other Mediterranean European countries. Robin W. Doughty, *The Eucalyptus: A Natural and Commercial History of the Gum Tree* (Baltimore: Johns Hopkins University Press, 2000), 43–46.

79. Following his first small, experimental planting of 62 trees, Cordier later planted at least 24,000 trees of some 120 different varieties of eucalyptus on his properties in the Mitidja. In the same region, Arlès-Dufour planted about 20,000 plants, which covered 20 hectares. Julien Franc, *La Colonisation de la Mitidja* (Paris: Honoré Champion, 1928), 637.

80. Eugène L. Bertherand, *L'Eucalyptus au point de vue de l'hygiène en Algérie* (Alger: Victor Aillaud, 1876), 15.

81. At this time miasmas, or bad smells, were believed by the European medical community to cause many diseases. It was not until the early 1880s that the germ theory of disease displaced the miasma theory in medical science.

82. Planchon, "L'Eucalyptus," 162.

83. Jules-Aimée Battandier and Louis Trabut, *L'Algérie: Le Sol et les habitants, flore, faune, géologie, anthropologie, ressources agricoles et économiques* (Paris: J.-B. Baillière, 1898), 105. The term "apostle" is theirs.

84. Ibid.

85. Bertherand, *L'Eucalyptus,* 34.

86. Achilles Filias, *Notice sur les forêts de l'Algérie: Leur étendue; leurs essences; leurs produits* (Alger: Gojosso, 1878), 20.

87. Bertherand, *L'Eucalyptus,* 9–10. The mining company Mokta-el-Hadid also planted many eucalyptus around Lake Fetzara in the 1870s, but all of their trees died because they were planted too close to the lake, which was highly saline. See Association Française pour l'Avancement des Sciences, *Compte rendu de la 10e session, Alger 1881* (Paris: Secrétariat de l'Association, 1882), 804–5.

88. Filias, *Notice,* 20. The army planted about 60,000 hectares of several different varieties of eucalyptus around Alger during the 1870s. See Association Française pour l'Avancement des Sciences, *Compte rendu,* 297–99.

89. Planchon, "L'Eucalyptus," 163.

90. Edward Pepper, "Eucalyptus in Algeria and Tunisia, from an Hygienic and Climatological Point of View," *Proceedings of the American Philosophical Society* 35, no. 150 (1896): 39–56.

91. Narcisse Faucon, *Le Livre d'or de l'Algérie: Histoire politique, militaire, administrative, événements et faits principaux, biographie des hommes ayant marqué dans l'armée, les sciences, les lettres, etc.* (Paris: Challamel, 1889), 516. He also won a gold medal for his eucalyptus plantations at the 1876 Algiers Exposition. There was even a street in Algiers named Trottier in the 1880s.

92. Franc, *La Colonisation,* 404, 512.

93. CAOM, ALG, GGA, P/59 and P/49.

94. François Trottier, *Notes sur l'eucalyptus et subsidiairement sur la nécessité du reboisement de l'Algérie* (Alger: F. Paysant, 1867), 5.

95. Ibid., 5–6.

96. Ibid., see 10–24.

97. Ibid., 11.

98. In Corsica, Trottier explained, an edict had been passed that forbade grazing in forests for five years after a fire. Ibid., 12.

99. François Trottier, *Boisement et colonisation: Rôle de l'eucalyptus en Algérie au point de vue des besoins locaux, de l'exportation, et du développement de la population* (Alger: Association Ouvrière, 1876).

100. Ibid., 24.

101. François Trottier, *Boisement dans le désert et colonisation* (Alger: F. Paysant, 1867). This work by Trottier is cited and discussed in Planchon, "L'Eucalyptus," 162–63, and Bertherand, *L'Eucalyptus,* 38.

102. Trottier, *Notes sur l'eucalyptus,* 22.

103. Trottier, *Boisement et colonisation,* 58.

104. Ibid., 92.

105. See, for example, the chapter by the director of the government's experimental garden at Algiers, Auguste Hardy, "Note climatologique sur l'Algérie au point de vue agricole," in *Recueil de traités d'agriculture et d'hygiène à l'usage des colons de l'Algérie,* ed. Ministre de la Guerre (Alger: Imprimerie du Gouvernement, 1851), 37–62.

106. See "Rapport au président de la république," Paris, 14 September 1875, and "Décret du président de la république française," 14 September 1875, CAOM, ALG, GGA, P/59. At this time Algeria was importing considerable amounts of wood for basic needs from Europe. See Filias, *Notice,* 43–44.

107. "Rapport au président de la république," Paris, 14 September 1875, and "Décret du président de la république française," 14 September 1875, CAOM, ALG, GGA, P/59.

108. Trottier, *Boisement et colonisation.*

109. Battandier and Trabut, *L'Algérie,* 106.

110. Pepper, "Eucalyptus," 48–49.

111. Franc, *La Colonisation,* 637. The one area where eucalyptus did have a beneficial impact was precisely in helping to dry out standing, stagnant water where the mosquitoes that transmit malaria preferred to breed. In this way they were "a precious auxiliary in cleaning up the swampy and poorly drained regions." Ibid. In the Mitidja region alone, 20,000 hectares of swampy land had been drained with hundreds of kilometers of canals by the 1930s. Although during the 1860s and 1870s eucalyptus alone (and its "beneficial vapor") was lauded as defeating malaria, actually it was this vast array of drainage canals constructed in swampy areas since the 1830s that was primarily responsible for the decline of malaria in Algeria. By reducing, and later eliminating, the breeding grounds for the *Anopheles* mosquito, the canals, aided in many areas by eucalyptus and other fast growing trees, helped to reduce malaria cases significantly from their previously high levels by the 1860s and 1870s. See Philip D. Curtin, *Death by Migration: Europe's Encounter with the Tropical World in the Nineteenth Century* (Cambridge: Cambridge University Press, 1989), 64–67. Although the *Anopheles* mosquito was not proven definitively to be the vector of malaria until 1898, the plasmodium that causes malaria was first microscopically identified in the blood of an Algerian soldier by a

military physician named Alphonse Laveran working in Bône in 1880. See Robert S. Desowitz, *The Malaria Capers: Tales of Parasites and People* (New York: W. W. Norton, 1991), 166–70. In the 1860s and 1870s (and for decades if not centuries before that) malaria was thought to be caused by "miasmas," that is, by smelly air in the vicinity of swamps. Physicians in Algeria continued to request the planting of eucalyptus groves as a way to improve the public health well into the 1880s. See Bertherand, *L'Eucalyptus*. It is a testament to the vast system of drainage canals constructed by the French during the colonial period (and to the judicious use of quinine) that the Maghreb is virtually free of malaria today.

112. See Association Française pour l'Avancement des Sciences, *Compte rendu*, 300.

113. A. D. Combe, *Les Forêts de l'Algérie* (Alger: Giralt, 1889).

114. Paul Boudy, *Économie forestière nord-africaine: Milieu physique et milieu humain*, vol. 1 (Paris: Larose, 1948), 467. Boudy does note that eucalyptus enjoyed more success in Morocco from the 1920s into the 1940s. Ibid., 467–68, see also xvii–xviii.

115. Trottier, *Boisement et colonisation*, 3. This argument is also evident in the archival documents on Trottier cited above.

116. Ibid., 64.

117. See Filias, *Notice*. This was a particularly important universal exposition for France since it was the first during the Third Republic and it showcased a reunified France following the Franco-Prussian War and the Commune.

118. Paulin Trolard was born in 1842 and completed his doctorate at the medical school in Paris in 1868. He emigrated to Algeria and by the early 1870s he was serving on the governing council of Algiers. He served on the governing council again in 1885 and likely served continuously for much of his time in Algeria. Trolard practiced medicine and taught at the medical school throughout his long career in Algeria. He married and raised several children. Although he hoped to retire in France, he died unexpectedly in 1910, at the age of 68, in Algeria. A prominent street was named after him in central Algiers. He published at least fifteen monographs in addition to articles in the *Bulletin*. For more details, see Ageron, *Les Algériens*, 1:50; *Bulletin de la Ligue du Reboisement de l'Algérie*, no. 18, n.s. (1910), 365; and Paulin Trolard, "Recherches sur l'anatomie du système veineux de l'encéphale et du crâne," thèse pour le doctorat en médecine, Faculté de Médecine de Paris, 1868.

119. *Bulletin de la Ligue du Reboisement de l'Algérie*, no. 1 (1882), 6. This is hereafter referred to as the *Bulletin de la Ligue*.

120. Ibid., 2, 6.

121. Ligue du Reboisement, *La Forêt: Conseils aux indigènes* (Alger: Association Ouvrière P. Fontana, 1883), 2.

122. See *Bulletin de la Ligue,* nos. 1–10 (1882). Of the roughly 1,200 members of the Ligue in November 1882, 46 were forest guards, 8 were mayors, 15 were in the military, 5 were forest subinspectors, 5 were forest inspectors, 3 were forest concessionaires, and 4 were newspaper editors. Other members included the directors of the Compagnie Algérienne and the Compagnie Genevoise, the prefect of Alger and the subprefect of Sétif, a judge, 7 caids, a police commissioner and a police inspector, elected members of municipal and general councils, professors of botany and medicine, attorneys, many property owners and farmers, as well as general business people. See *Bulletin de la Ligue,* no. 7 (1882), supplement, and other 1882 issues.

123. In the early years, the bulk of the members were foresters. See Hippolyte Doumenjou, *Études sur la révision du code forestier: Les Reboisements en France et en Algérie* (Paris: J. Baudry, 1883), 265. Other important founding members included the director of the Jardin d'Essai, M. Rivière, attorney H. Lecq, and veterinarian E. Bonzom. Ernest Carette joined the Ligue in 1882.

124. The position of forest conservator was at the time the highest in the forestry department. M. Mangin was chief forest conservator from 1873 to 1883. Mangin had also been a member of the Society for the Encouragement of Reforestation in Algeria, founded in 1874. Beginning in 1881, each of the three provinces was assigned its own conservator. In 1898 another reorganization resulted in a single chief forest inspector for all of Algeria. Henri Lefebvre served in this capacity from 1898 to 1903 and V. Boutilly served from 1903 to 1910. Boutilly was also a member of the Ligue.

125. For examples of these tensions and discussion of the issues, see Ageron, *Les Algériens,* 1:113; and Louis-François Victor Tassy, *Service forestier de l'Algérie, rapport adressé à m. le gouverneur de l'Algérie* (Paris: A. Hennuyer, 1872), 16–24.

126. Ageron, *Les Algériens,* 1:127–28. Corvée labor was even imposed on Algerians during this period as punishment for forest crimes. Ibid., 1:113.

127. Commandant P. Wachi, *La Question des forêts en Afrique* (Tunis: Imprimerie Rapide, 1899), 6.

128. Ibid., 11.

129. Tassy, *Service.* See Boudy, *Économie forestière,* 1:242–46, for a discussion of the influence of Tassy's report.

130. Martine Chalvet, "Les Enjeux de la politique forestière en Algérie," *Ultramarines: Bulletin des Amis des Archives d'Outre-Mer,* no. 4 (December 1991): 14.

131. *Bulletin de la Ligue,* no. 1 (1882), 2, 6.

132. *Bulletin de la Ligue,* no. 2/3 (1882), 22. These letters, written in support of the Ligue, came from government functionaries such as the prefect of Algiers and the subprefect of Bougie, as well as the presidents of the chamber of commerce and the comice agricole. Even settlers and soldiers wrote him with enthusiasm. The Ligue's efforts to raise public awareness in Algeria, and the stir created in the local press, were even noted by politicians in France. See Doumenjou, *Études,* 265.

133. For many examples of the letters and petitions to all levels of government from the Ligue over the years see CAOM, ALG, GGA, P/59, folder on "Reboisements, Ligue du Reboisement."

134. *Bulletin de la Ligue,* no. 2/3 (1882), 20.

135. See CAOM, ALG, GGA, P/59, folder on reboisement.

136. Tirman, "Governor's Report," 1884, reprinted in *Bulletin de la Ligue,* no. 97 (1889), 1781.

137. The Ligue's influence was noted, with approval, by a legal historian in 1906. See Angel-Paul Carayol, *La Législation forestière de l'Algérie* (Paris: Arthur Rousseau, 1906), 167.

138. *Bulletin de la Ligue,* no. 2/3 (1882), 23.

139. Statement by M. Bourlier made in the Superior Council, quoted in *Bulletin de la Ligue,* no. 14 (1883), 279.

140. See Letter of 29 March 1883 from the Ministère de l'Agriculture, Direction des Forêts, Paris, to M. le Lievre, Senateur de Department d'Alger, reprinted in *Bulletin de la Ligue,* no. 18 (1883), 345–46. Encouraged by de Mahy's earlier support of reforestation on the Île Bourbon (Réunion), Trolard correctly believed that he would also support reforestation efforts in Algeria.

141. CAOM, ALG, GGA, P/59, folder on the Ligue du Reboisement, dossier on Trolard's petition for recognition of public utility.

142. *Bulletin de la Ligue,* no. 256 (1886), 1090–91.

143. In one of these cases, the prefect, a member of the Ligue, called in the authorities to deal with "this act of vandalism," while in another the prefect initiated an investigation into the "marauding natives." *Bulletin de la Ligue,* no. 2/3 (1882), 23–34.

144. See CAOM, ALG, GGA, P/59 and P/49, for numerous examples. The government appears to have kept a close watch on the Ligue and archived many Ligue documents and publications in its files.

145. *Bulletin de la Ligue,* no. 7 (1882), 132–35.

146. *Bulletin de la Ligue,* no. 15 (1883), 296.

147. *Bulletin de la Ligue,* no. 53 (1886), 1055.

148. Gouvernement Général de l'Algérie, *Programme générale du reboise-ment* (Alger: Imprimerie Administrative, 1885), 9.

149. *Bulletin de la Ligue,* no. 48 (1885), 966.

150. The 1885 law was presented to the French parliament, and adopted very quickly with little debate, following the report of a new commission that included Eugène Étienne, an activist for forest protection. Étienne's motives appear to have been more economic than ecological, however. He wrote in his report on the new law that "without the contraction of use rights and the suppression of cultivated enclaves [in the forest] there will not be industrial exploitation of the forests." Ageron, *Les Algériens,* 1:124.

151. Carayol, *La Législation,* 55. See 31–58 for details of the 1885 law.

152. See Gouvernement Général de l'Algérie, *Programme,* 6.

153. Ibid., 13–14, 37, 47, and 67.

154. Ibid., 59–60.

155. Ibid., 112.

156. Ibid., 35–36. Bert praises the work of the Ligue in this concluding section of his report.

157. Combe, *Les Forêts,* 57.

158. Gouvernement Général de l'Algérie, *Programme,* 41.

159. Combe, *Les Forêts,* 9.

160. Gouvernement Général de l'Algérie, *Programme,* 69.

161. Ibid., 8.

162. Ibid., 47.

163. A. D. Combe, "Notice sur les forêts de l'Algérie," in *Catalogue des collections exposées à l'Exposition Universelle et Internationale,* ed. Service des Forêts (Alger: Adolphe Jourdan, 1885), 3–4.

164. Ibid., 4.

165. Ibid., 39.

166. Ibid., 36.

167. Ibid., 4. An 1875 decree had placed about three-quarters of a million hectares of "forest land" under the control of the military, most of it in the south. See V. Boutilly, *Recueil de la législation forestière algérienne, lois, décrets et règlements divers* (Paris: Berger-Levrault, 1905), 186–87. The Forest Service wanted to have this decree rescinded in order to regain control over these southern forests. Gouvernement Général de l'Algérie, *Programme,* 60.

168. J. Reynard, *Restauration des forêts et des pâturages du sud de l'Algérie (province d'Alger)* (Alger: Adolphe Jourdan, 1880), 6.

169. Ibid., 15–16.

170. Ibid., 38.

171. Ibid., 36.

172. Ibid., 35.

173. Ibid., 7.

174. See Tassy, *Service*, 5, and Combe, *Les Forêts*, 9.

175. See Combe, *Les Forêts*, 12–13, for a statistical breakdown of forest holdings at this time.

176. Senator Guichard, quoted in Ageron, *Les Algériens*, 1:126.

177. Ibid.

178. Charles-Robert Ageron, *Histoire de l'Algérie contemporaine: De l'insurrection de 1871 au déclenchement de la guerre de libération (1954)* (Paris: Presses Universitaires de France, 1979), 208.

179. *Bulletin de la Ligue*, no. 97 (1889), 1784.

180. Ageron, *Les Algériens*, 1:128.

181. See Boudy *Économie forestière*, 1:249–52, for a typical evaluation of the Ferry report by foresters.

182. See *Bulletin de la Ligue* for the years 1890–94.

183. See Carayol, *La Législation*, 83–88, for the complicated history of the development and adoption of the 1903 Algerian forest law. The 1893 *projet de loi* itself contained typical desiccationist arguments and assumptions about deforestation leading inevitably to desertification. See Gouvernement Général de l'Algérie, *Commission de législation forestière de l'Algérie: Projet de loi* (Alger: Giralt, 1893), available at CAOM.

184. See Henri Lefebvre, *Les Forêts de l'Algérie* (Alger-Mustapha: Giralt, 1900), 99–106. Lefebvre had served in Algeria before becoming the director of the Forest Service in Tunisia in 1883. He served in that capacity until returning to Algeria as chief forest inspector and head of the new "technical forestry service" created in 1898. See Boudy, *Économie forestière*, 1:252, 262, for details.

185. Lefebvre, *Les Forêts*, 101.

186. Ibid., 118–19.

187. Ibid., 117–18.

188. *Bulletin de la Ligue*, no. 111 (1894), 2309.

189. Tassy, *Service*, 39–40.

190. Ageron, *Histoire*, 208.

191. Reprinted in the *Bulletin de la Ligue*, no. 119 (1898), 2523.

192. Paulin Trolard, *La Colonisation et la question forestière* (Alger: Casabianca, 1891), 77.

193. Ibid., 15.

194. Ageron, *Les Algériens*, 2:776. For good overviews of the 1903 law and detailed discussions, see Boutilly, *Recueil;* Carayol, *La Législation;* and Theodore

Woolsey, *French Forests and Forestry: Tunisia, Algeria, Corsica* (New York: John Wiley, 1917). Both Boutilly and Woolsey provide the complete text of the law.

195. Carayol, *La Législation,* 153.

196. Ibid., 55.

197. Ageron, *Les Algériens,* 1:124.

198. See Carayol, *La Législation,* 169–71, for details on the reforestation perimeters.

199. Ibid.

200. Boudy, *Économie forestière,* 1:479.

201. Paul Boudy, *Économie forestière nord-africaine: Description forestière de l'Algérie et de la Tunisie,* vol. 4 (Paris: Larose, 1955), 380–82. The Special Reforestation Service was first directed by Paul Boudy, who later directed the forestry service in Morocco for most of the colonial period. The Reforestation Service of Algeria mutated into the Service for the Defense and Restoration of Soils (DRS) in 1941.

202. See ibid. for the statistics on perimeters and amount of land "reforested."

203. Ageron, *Les Algériens,* 2:781.

204. Gouvernement Général de l'Algérie, *Commission d'études forestières* (Alger: J. Torrent, 1904), 5.

205. See Ageron, *Les Algériens,* 2:779–80.

206. See Gouvernement Général de l'Algérie, *Commission d'études,* 55–118, for Trolard's two reports. These comprise about half (64 of 131 pages) of the entire report exclusive of the annexes. Ten other men were members of the commission but in a lesser capacity than the ten reporting members.

207. Ibid., 101.

208. *Bulletin de la Ligue,* no. 6, n.s. (1905), 119.

209. Ibid.

210. Jonnart, quoted ibid., 116.

211. Dessoliers, quoted in *Bulletin de la Ligue,* no. 8, n.s. (1906), 133.

212. See chapters 2 and 3 above for details of early botanical work on Algeria.

213. Only the major, influential botanists are discussed here, not the many amateurs, including colonists, physicians, army personnel, and others, who occasionally wrote notes on plants they found in the course of their work.

214. For biographical details on Trabut see René Maire, "Louis Trabut," *Revue de Botanique Appliquée et d'Agriculture Tropicale* 9, no. 98 (1929): 613–20; and René Maire, *Les Progrés des connaissances botaniques en Algérie depuis 1830* (Paris: Masson, 1931), 137–45.

215. For more details on Trabut's life and work, see Maire, "Louis Trabut."

216. For biographical details on Battandier, see Maire, *Les Progrès;* and René Maire and Charles-Louis Trabut, "Jules-Aimé Battandier (1848–1922)," *Bulletin de la Société Botanique de France* 23, no. 23 (February 1923): 106–17.

217. René Maire, *Flore de l'Afrique du nord (Maroc, Algérie, Tunisie, Tripolitaine, Cyrénaique et Sahara)* (Paris: Paul Lechevalier, 1952).

218. Auguste Mathieu and Louis Trabut, *Les Hauts-plateaux Oranais, rapport de mission* (Alger: Pierre Fontana, 1891).

219. Ibid., 14–15.

220. See Reynard, *Restauration.*

221. Mathieu, *Les Hauts-plateaux,* 42.

222. Ibid., 46–47.

223. Alfa grass (sometimes spelled halfa) is the perennial grass *Stipa tenacissima.* It was harvested in North Africa and in Spain during the nineteenth century for a variety of industrial purposes, the most important of which was making paper.

224. Ibid., 29. It is worth noting that Mathieu and Trabut thought this region inappropriate for colonization by Europeans, and this likely influenced their conclusions.

225. Battandier and Trabut, *L'Algérie.*

226. Ibid., 18.

227. Ibid., 24.

228. Ibid., 197.

229. Ibid., 355. The Quaternary period is the current period of geologic time, which began about one million years ago. It includes the last Ice Age (Pleistocene) and the more recent Holocene epoch. At the time this was written, the dates were not this precise.

230. Ibid.

231. Ibid., 15.

232. Ibid., 147–48.

233. It is interesting to note, however, that nearly twenty years earlier Trabut had published an article that stated that the High Plateaus could only be "conquered for colonization by afforestation which, in modifying the climate, will bring to the soil the humidity that the prairies need." See Louis Trabut, "Les Régions botaniques et agricoles de l'Algérie," *La Revue Scientifique* 1, no. 15 (1881): 464–65. This article displays other desiccationist sentiments that are not found in his later writings. Trabut was one of the early members of the Ligue du Reboisement, a fact that may have influenced his writing at that time. Ligue member Reynard's work on the High Plateaus (*Restauration des forêts*) had appeared the year before and there are many

similarities to it in the section Trabut wrote on the High Plateaus in his 1881 article.

234. Battandier and Trabut, *L'Algérie*, 213.

235. Ibid., 219.

236. Ibid., v.

237. See, for some examples, Cosson, *Compendium;* Edmond Lefranc, "Sidi-Bel-Abbès: Topographie, Climatologie et Botanique," *Bulletin de la Société Botanique de France* 12 (1865): 383–95 and 415–31; Svante S. Murbeck, *Contributions à la connaissance de la flore du nord-ouest de l'Afrique et plus spécialement de la Tunisie* (Lund: E. Malmstrom, 1897); Svante S. Murbeck, *Contributions à la connaissance de la flore du nord-ouest de l'Afrique et plus spécialement de la Tunisie, deuxième série* (Lund: Hakan Ohlsson, 1905); l'Abbé Chevalier, "Deuxième note sur la flore du Sahara," *Bulletin de l'Herbier Boissier* 3, no. 1 (1903): 669–84.

238. See *Bulletin de la Ligue,* no. 85 (1888), 1601–4; no. 86 (1889), 1609–13; no. 87 (1889), 1629–31.

239. Phytogeography, in turn, would develop into phytosociology (plant ecology) in the early twentieth century and have profound implications for the environmental history of North Africa, as the next chapter details.

240. Ernest Cosson, *Le Règne végétal en Algérie* (Paris: A. Quantin, 1879).

241. Trabut, "Les Régions," reprinted in *Bulletin de la Ligue.* See note 238, above.

242. Ibid., 1602.

243. Ibid., 1603.

244. Ibid., 1610–11.

245. See, for example, E. Ficheur, "Relation entre la constitution géologique du sol et la distribution des boisements," in *Les Forêts de l'Algérie,* ed. Henri Lefebvre (Alger-Mustapha: Giralt, 1900), 70–91.

Chapter 5

1. Charles-Robert Ageron, *Modern Algeria: A History from 1830 to the Present,* trans. Michael Brett (London: Hearst, 1991), 61, 83.

2. See Charles-Robert Ageron, *Histoire de l'Algérie contemporaine: De l'insurrection de 1871 au déclenchement de la guerre de libération (1954)* (Paris: Presses Universitaires de France, 1979), 481, for details of this 1926 property law.

3. Ageron, *Modern,* 61.

4. Ibid., 88.

5. Ibid.

6. Ibid., 89.

7. Ibid., 90.

8. See Jamil M. Abun-Nasr, *A History of the Maghrib* (Cambridge: Cambridge University Press, 1971), for more details of Tunisia's history. See also Kenneth J. Perkins, *A History of Modern Tunisia* (New York: Cambridge University Press, 2004).

9. Abun-Nasr, *History*, 281–82.

10. Ibid.

11. Ibid., 343–44.

12. Jean Poncet and André Raymond, *La Tunisie* (Paris: Presses Universitaires de France, 1971), 33.

13. Abun-Nasr, *History*, 344.

14. Poncet and Raymond, *La Tunisie*, 33–34.

15. Ibid., 34.

16. Henri Lefebvre, *Notice sur les forêts de la Tunisie et catalogue raisonné* (Tunis: B. Borrel, 1889), provides details of Tunisian forests and the Forest Service to 1889. Of the 810,746 hectares of "forest land," nearly 160,000 hectares (roughly 20 percent) was actually unforested land or scrubland. Ibid., 18.

17. Ibid., 33.

18. Ibid., 36–37.

19. Ibid.

20. See Résidence Générale de France à Tunis, Service des Forêts, *Recueil des décrets et arrétés constituant la législation forestière tunisienne* (Tunis: Imprimerie Centrale, 1936), for details.

21. See Habib Attia, "L'Évolution des structures sociales et économiques dans les hautes steppes," *Revue Tunisienne des Sciences Sociales* 3, no. 6 (1966): 5–38, for a vivid and compelling description of this process in the High Steppes of central Tunisia. See also Poncet, *La Tunisie*, 31–34, for more details on property law during the colonial period.

22. Résidence Générale de France à Tunis, Service des Affaires Indigènes, *Historique du bureau des affaires indigènes de Matmata* (Bourg: Victor Berthod, 1931), 23.

23. Résidence Générale de France à Tunis, Service des Affaires Indigènes, *Historique de l'annexe des affaires indigènes de Ben-Gardane* (Bourg: Victor Berthod, 1931), 13.

24. For information on Bourde's interesting life, see Anon., *La Vie studieuse, féconde et désintéressée de Paul Bourde, publiciste colonial, agronome, rénovateur des oliveraies tunisiennes* (Paris: Montsouris, 1943). Born in 1851 in France, Bourde became a journalist, the capacity in which he first visited Tunisia in

1881. He returned in 1888 and within a few years he had been appointed director of agriculture by the resident general of Tunisia. In 1895 he went to Madagascar to serve as an adjunct to the governor. Within the year, however, he retired, due to ill health, to France, where he died fifteen years later.

25. For details on this law, see Paul Bourde, *Rapport sur les cultures fruitières et en particulier sur la culture de l'olivier dans le centre de la Tunisie* (Tunis: Imprimerie Générale [Picard], 1899), 65–68.

26. Ibid., 3.

27. Ibid., 5.

28. Jean Poncet, *Paysages et problèmes ruraux en Tunisie* (Paris: Presses Universitaires de France, 1962), 257–58. Poncet believes that, early in this process, many of those planting olive groves were indigenous Tunisians who lost their holdings later, and thus not necessarily due to the 1892 decree. See 259–60. I am grateful to Paul Claval for first telling me about the works of Paul Bourde.

29. For the complex historical details of the conquest of Morocco, see Edmund Burke, *Prelude to Protectorate in Morocco: Precolonial Protest and Resistance, 1860–1912* (Chicago: University of Chicago Press, 1976). Morocco had not been under Ottoman rule as Algeria and Tunisia had been.

30. Louis Lyautey, *Paroles d'action* (Paris: Imprimerie Nationale, 1995), 444–45.

31. The Greek myth of the Garden of the Hesperides told of a tree that bore golden apples and figured prominently in the myth of Hercules. Some speculated that the Garden of the Hesperides was in Arcadia, thus making it easy to conflate Morocco and the Arcadian ideal.

32. For more details on, and an excellent discussion of, the influence of the more positive aspects of the narrative in the agricultural policies of the northwestern coastal plains, see Will D. Swearingen, *Moroccan Mirages: Agrarian Dreams and Deceptions, 1912–1986* (Princeton, N.J.: Princeton University Press, 1987). This area of the northwestern coastal plains was known as "useful Morocco" (*Maroc utile*), while the areas to the east and the south of the Atlas Mountains were thought to be less agriculturally productive and thus less useful to the colonial project.

33. Jean Brignon et al., *Histoire du Maroc* (Paris: Hatier, 1967), 354, 378.

34. Daniel Rivet, *Lyautey et l'instauration du protectorat français au Maroc 1912–1925* (Paris: Harmattan, 1988), 2:206–7.

35. Ibid. Rivet notes that the 1919 law was inspired by similar legislation in Tunisia that had forced many Tunisians to sell their land, which was then made available for colonization. See 2:208.

36. Paul Boudy, *Économie forestière nord-africaine: Milieu physique et milieu humain*, vol. 1 (Paris: Larose, 1948), 306–8. The amount of land defined as forest land grew during the colonial period to the point that by the end of the protectorate period about five million hectares were so classified.

37. Brignon et al., *Histoire*, 373, 378.

38. For an excellent analysis of resistance in the south, see Ross E. Dunn, *Resistance in the Desert: Moroccan Responses to French Imperialism, 1881–1912* (London: Croom Helm, 1977).

39. Rivet, *Lyautey*, 2:202.

40. P. Christian, *L'Afrique française, l'empire de Maroc et les déserts de Sahara* (Paris: A. Barbier, 1846), 52. The ancient Greeks' pillars of Hercules have long been accepted as being the mountains on either side of the Straits of Gibraltar (in Spain and Morocco); Hercules, pushing them apart, created the Straits.

41. Xavier Durrieu, *The Present State of Morocco: A Chapter of Mussulman Civilisation* (London: Longman, Brown, Green, and Longmans, 1854), 49.

42. Maurice Besnier, "La Géographie économique du Maroc dans l'antiquité," *Archives Marocaines* 7, no. 1 (1906): 276.

43. Maurice Besnier, *Géographie ancienne du Maroc* (Paris: Ernest Leroux, 1904), 50. See also Besnier, "Géographie Économique," 271. For a good discussion of the government-sanctioned Mission Scientifique du Maroc, see Edmund Burke, "La Mission scientifique au Maroc," in *Actes de Durham: Recherches récentes sur le Maroc moderne*, ed. P. Pascon (Rabat: B.E.S.M., 1979), 37–56. *The Scientific Exploration of Morocco* (*Exploration Scientifique du Maroc*), a later, separate effort, also sanctioned by the French government but organized by the Geographical Society, produced a botanical survey in 1913, in what was likely the first such survey of the protectorate. Although this survey is nearly entirely a long list cataloging the plants of western Morocco, it does contain traces of the narrative in the preface. See M. C.-J. Pitard, *Exploration scientifique du Maroc: Premier fascicule: Botanique* (Paris: Masson, 1913). This exploratory mission also included studies of the agronomy and zoology of the protectorate.

44. Jean Célérier, *Histoire du Maroc: Sommaire des leçons professées au cours de perfectionnement du service des renseignements* (Rabat: Direction des Affaires Indigènes et du Service des Renseignements, 1921), 14.

45. Auguste Terrier, *Le Maroc* (Paris: Larousse, 1931), 187; and Stéphane Gsell, *Histoire ancienne de l'Afrique du nord: Les Conditions du développement historique, les temps primitifs, la colonisation phénicienne et l'empire de Carthage* (Paris: Hachette, 1921).

46. Paul Boudy, "L'Oeuvre forestière française au Maroc," *Bulletin de la So-ciété Forestière de Franche-Comté et des Provinces de l'Est* 27, no. 1 (1954): 2. For additional information on Boudy, including biographical details, see Philibert Guinier, "Nécrologie: Paul Boudy (1874–1957)," *Revue Forestière Française* 3 (May 1958): 219–22.

47. Between 1936 and 1939, for example, Boudy was put in charge of reor-ganizing the important property and cadastral service. See Guinier, "Nécrolo-gie," 220. Boudy also worked closely with Resident General Lyautey on many issues, including, but not limited to, forestry. He did, though, oppose Lyautey on at least one occasion by strongly advising against giving a concession for exclusive private exploitation of the Ainleuh Forest to a Frenchman, M. de Nuchèze. See telegram of 9 March 1922 from Urbain Blanc, Délegué of the Resident General Lyautey, to Lyautey, Maroc/DAI/Inv. 8/319a, Centre des Archives Diplomatiques de Nantes (hereafter CADN).

48. Guinier, "Nécrologie," 219. Guinier thought so highly of Boudy and his oeuvre that he wrote that his work "recalls, with even more amplitude, that of Demontzey," the cherished father of reforestation in France. Ibid., 222.

49. For information on Boudy in Morocco, see letter no. 290 of 23 March 1912, Maroc/dossiers personnels, carton two, CADN, and letter no. 141 of 3 March 1913, Maroc/dossiers personnels, carton two, CADN.

50. Général Lyautey, ed. *Rapport général sur la situation du protectorat du Maroc* (Rabat: Résidence Générale de la République Française au Maroc, 1916).

51. Henri Marc, *Notes sur les forêts de l'Algérie* (Alger: Adolphe Jourdan, 1916).

52. Boudy, *Économie forestière*, 1:274.

53. Paul Boudy, "Les Forêts du Maroc," *La Revue Générale des Sciences Pures et Appliquées* 25, no. 7 (1914): 350–54.

54. For some of the details of the early inspections and decisions regarding Morocco's forests, see "Service des Forêts," no date, in Maroc/DAI/Inv. 8/#101, CADN. The French geographer Jean-Yves Puyo notes that Lyautey also derived some inspiration for the creation of the Moroccan Forest Service from French Indochina. He explains, furthermore, to what extent economic factors influ-enced the impetus and organization of the service. Jean-Yves Puyo, "Lyautey et la politique forestière marocaine (Protectorat français, 1912–1956)." Paper pre-sented at Géographie, Exploration et Colonisation (XIXe–XXe siècles), Col-loque Bordeaux, 13–15 October 2005.

55. Daniel Rivet, however, makes clear how much violence actually was used during the military campaigns of pacification in Morocco. See Rivet, *Lyautey*.

56. Lyautey, *Rapport,* 419.

57. Boudy, *Économie forestière,* 1:306–8.

58. See "Dahir sur l'exploitation et conservation des forêts," 1917, Maroc /DAI/Inv. 8/319a, CADN, for more details on the Moroccan Forest Code.

59. See ibid.

60. In the south near Essaouira and Agadir, for example, the Forest Service did not establish a presence until 1920. It was not until 1925 that the law was decreed to govern the delimitation and protection of the argan forests in the region, bringing them under the 1917 Moroccan Forest Code. Boudy, *Économie forestière,* 1:276–77.

61. André Metro, *Forêts, Atlas du Maroc, notices explicatives, section VI, biogéographie: Forêts et ressources végétales* (Rabat: Comité de Géographie du Maroc, 1958).

62. For information on the history of phytogeography, see Patrick Matagne, *Aux Origines de l'écologie: Les Naturalistes en France de 1800 à 1914* (Paris: CTHS, 1999), 85–88.

63. Buffon had earlier pointed out the importance of the physical properties of the environment in the distribution of plants around the globe.

64. Janet Browne, *The Secular Ark: Studies in the History of Biogeography* (New Haven, Conn.: Yale University Press, 1983), 48–57. There was active debate over nomenclature in botanical geography, and later phytogeography, in the late nineteenth and early twentieth centuries. Terms such as associations, regions, formations, stations, zones, etc. were often used loosely until about World War I.

65. See Augustin de Candolle, "Géographie botanique," in *Ecological Phytogeography in the Nineteenth Century,* ed. Frank N. Egerton (New York: Arno, 1977), 422–37. Humboldt also published voluminously in other venues.

66. For information on Alphonse de Candolle and his influence on botanical geography, see Browne, *Secular Ark;* and Matagne, *Aux Origines.*

67. Matagne, *Aux Origines,* 88.

68. Ibid., 223–37. His students included Max Sorre and phytosociologist Braun-Blanquet, among others.

69. Charles Flahault, "Au Sujet de la carte botanique, forestière et agricole de France," *Annales de Géographie* 5, no. 15 (October 1896): 456.

70. Ibid.

71. For more information, including contemporary phytosociological methods, see Michael G. Barbour et al., *Terrestrial Plant Ecology,* 3rd ed. (Menlo Park, Calif.: Addison, Wesley, Longman, 1999), 213–17, 241–44. For an early critique of this method, see Frank E. Egler, "Philosophical and Practical

Considerations of the Braun-Blanquet System of Phytosociology," *Castanea* 19, no. 2 (1954): 45–60. Many other developments in geographic botany and plant ecology took place in Europe and North America during the early twentieth century. For an overview, see Hugh M. Raup, "Trends in the Development of Geographic Botany," *Annals of the Association of American Geographers* 32, no. 4 (1942): 319–54.

72. For details on Flahault's life and oeuvre, see Louis Emberger, "Charles Flahault (1852–1935)," *Revue Générale de Botanique* 48, no. 565 (1936): 1–48.

73. See *Bulletin de la Société Botanique de France* (1906), tome 53, i–clxv.

74. For biographical information on Maire, see Louis Emberger, "Préface," in *Flore de l'Afrique du nord,* ed. René Maire (Paris: Paul Lechevalier, 1952), 1–6; and Philibert Guinier, "René Maire, 1878–1949, In Memoriam," *Bulletin de la Société d'Histoire Naturelle de l'Afrique du Nord* 41, no. 1 (1950): 64–113.

75. Two other men collaborated on the *Carte Phytogéographique de l'Algérie-Tunisie:* Paul de Peyerimhoff and G. Lapie, both of them foresters. Maire, a prodigious researcher, published more than three hundred books and articles (of which about one-third dealt with mycology, the study of fungi).

76. Guinier, "René Maire," 71, 102–5. For a touching homage to Maire by Emberger, see Emberger, "Préface."

77. René Maire, "Coup d'oeil sur la végétation du Maroc," in *Sur les productions végétales du Maroc,* ed. Émile Perrot and Louis Gentil (Paris: Larose, 1921), 59–71.

78. Ibid., 60.

79. Ibid., 69. In doing this, Maire and others may have been inspired by the example of Plato, who looked to similar sanctuaries of his day for what he considered the vestiges of formerly forested areas around Athens. See Clarence Glacken, *Traces on the Rhodian Shore* (Berkeley: University of California Press, 1967), 121.

80. Maire, "Coup d'oeil," 62.

81. Ibid.

82. See H. F. Lamb, U. Eichner, and V. R. Switsur, "An 18,000-Year Record of Vegetation, Lake-Level and Climate Change from Tigalmamine, Middle Atlas, Morocco," *Journal of Biogeography* 1, no. 16 (1989): 65–74; and H. F. Lamb, F. Damblon, and R. W. Maxted, "Human Impact on the Vegetation of the Middle Atlas, Morocco, During the Last 5000 Years," *Journal of Biogeography* 18, no. 5 (1991): 519–32.

83. Nor did Maire invoke the declensionist environmental narrative in his publications on Algeria. He maintained the more restrained tone of the earlier Algerian botanists in this regard. See, in particular, his government-

commissioned publication, René Maire, *Carte phytogéographique de l'Algérie et de la Tunisie: Notice* (Alger: Baconnier Frères, 1926).

84. Paul Boudy, *L'Arbre et les forêts au Maroc. Cours préparatoire au service des affaires indigènes* (Rabat: Résidence Générale de France au Maroc, Direction Générale des Affaires Indigènes, 1927), 8.

85. During his second sojourn in Algeria, Boudy joined the Ligue du Reboisement and was likely influenced by its rhetoric, which is echoed in much of his writing. He praised Trolard and the Ligue in his writings. See Boudy, *Économie forestière,* 1:250.

86. Boudy, *L'Arbre,* 17.

87. Ibid.

88. Ibid., 10.

89. Ibid., 21–22. The concept of the *taux de boisement* and its importance may be traced back to at least the nineteenth century in France and likely much earlier. A taux de boisement of 30 percent was the widely accepted minimum for civilization in France and much of Europe. It was applied to Algeria to make much the same argument in favor of extensive reforestation, by Trolard and others, as detailed in the last chapter.

90. Paul Boudy and Louis Emberger, "La Forêt marocaine," in *La Science au Maroc,* ed. Paul Boudy and P. Despujols (Casablanca: Imprimeries Réunies, 1934), 191.

91. Louis Emberger, "Aperçu général [de la végétation]," in *La Science au Maroc,* ed. Paul Boudy and P. Despujols (Casablanca: Imprimeries Réunies, 1934), 163. At that time Morocco was thought to have about 50 million hectares within its borders. See Jean Célérier, *Le Maroc* (Paris: Armand Colin, 1931), 7. Thus, Emberger believed at this time that at least 40 percent of the entire territory should be forested, and had been "naturally" forested in the past.

92. Emberger, "Aperçu général [de la végétation]," 163.

93. For details on Emberger's life, see Jacques Miège, "In Memoriam: Louis Emberger," *Candollea* 25, no. 2 (1970): 183–87.

94. Emberger, quoted in G. Mangenot, "La Carrière et l'oeuvre," in *Travaux de botanique et d'écologie,* by Louis Emberger (Paris: Masson, 1971), 2.

95. The original Institute of Botany was founded by Flahault in 1889–90. The former CEPE is currently named the Centre d'Écologie Evolutive et Fonctionelle (CEFE).

96. Imanuel Noy-Meir and Eddy van der Maarel, "Relations between Community Theory and Community Analysis in Vegetation Science: Some Historical Perspectives," *Vegetatio* 69, no. 1–3 (1987): 7.

97. I. Nahal, "The Mediterranean Climate from a Biological Viewpoint," in *Mediterranean-Type Shrublands,* ed. Francesco di Castri, David W. Goodall, and Raymond Specht (New York: Elsevier Scientific, 1981), 63–86.

98. Louis Emberger, "La Végétation de la région méditerranéenne: Essai d'une classification des groupements végétaux," *Revue Générale de Botanique* 42, no. 503 (1930): 718. Emberger published many articles on the Mediterranean basin and arid lands more generally. Much of this research was based on, and informed by, his Moroccan work.

99. Ibid., 708.

100. Louis Emberger, "Aperçu général sur la végétation du Maroc. Commentaire de la carte phytogéographique du Maroc," *Veröffentlichungen des Geobotanischen Institutes Rübel in Zürich* 14, no. 1 (1939): 43.

101. Marvin Mikesell, "Deforestation in Northern Morocco," *Science* 132, no. 3425 (1960): 441–48.

102. For a discussion of the serious problems involved in deducing natural vegetation from relict vegetation elsewhere in Africa, see James Fairhead and Melissa Leach, *Misreading the African Landscape: Society and Ecology in a Forest-Savanna Mosaic* (Cambridge: Cambridge University Press, 1996); and James Fairhead and Melissa Leach, *Reframing Deforestation: Global Analyses and Local Realities; Studies in West Africa* (London: Routledge, 1998).

103. See chapter 1 above for details.

104. See for example, Douglas G. Sprugel, "Disturbance, Equilibrium, and Environmental Variability: What is 'Natural' Vegetation in a Changing Environment?" *Biological Conservation* 58, no. 1 (1991): 1–18.

105. Emberger, "Aperçu général sur la végétation du Maroc," 44. Emberger referenced several of Flahault's publications in this work. He specifically cited Flahault's map of the botanical, forest, and agricultural regions of France as providing the general principles guiding his phytogeographic map of Morocco. See ibid., 43.

106. Ibid, 43.

107. Emberger, "Aperçu général [de la végétation]," 163.

108. Emberger, "Aperçu général sur la végétation du Maroc," 45.

109. Ibid., 40–157. This beautiful map is available as a reprint in the volume dedicated to Emberger's oeuvre published shortly after his death. See Louis Emberger, *Travaux de botanique et d'écologie* (Paris: Masson, 1971).

110. Charles Sauvage, "La Forêt marocaine," *Bulletin de l'Enseignement Publique* 28, no. 169 (1941): 250.

111. Boudy, *Économie forestière,"* 1:170.

112. Ibid., 1:215. Boudy makes this calculation slightly more clear in his 1958 volume in the series dedicated to Morocco. See Paul Boudy, *Économie forestière nord-africaine: Description forestière du Maroc,* vol. 3 (Paris: Larose, 1958), 351. Elsewhere in the 1948 volume, however, in a section perhaps written earlier, he estimated that Morocco was 66 percent deforested, but did not provide an ecological calculation. See Boudy, *Économie forestière,* 1:267. This may be a repetition of what he estimated in 1927. See Boudy, *L'Arbre,* 22.

113. Boudy also calculated higher deforestation rates when he included Moroccan "scrub" (considered degraded forest), and thus attained between 65 and 69 percent deforestation rates. Boudy, *Économie forestière,* 1:215.

114. Ibid., 1:169.

115. Ibid., 1:170.

116. It is interesting that this is the exact estimate of deforestation for Algeria that Paulin Trolard, founder of the Ligue du Reboisement de l'Algérie, had made in 1904. See Gouvernement Général de l'Algérie, *Commission d'études forestières* (Alger: J. Torrent, 1904), 113.

117. Boudy, *Économie forestière,* 1:227–28. He also highlighted additional causes of deforestation in Morocco, including the construction of great cities such as Fez and Marrakech, the tanning industry, and the presumed destructiveness of the numerous goats.

118. For representative statements on the evils of grazing throughout the colonial period, see Anon., "La Forêt marocaine," in *Études marocaines: Supplément au bulletin d'informations et de documentation de la résidence générale* (Rabat: Résidence Générale de France au Maroc, 1940), 5–28; Augustin Bernard, *Une Mission au Maroc* (Paris: Le Comité du Maroc, 1904); Louis Emberger, "L'Importance du chêne-liège dans le paysage marocain avant la destruction des forêts," *La Revue de Géographie Marocaine* 7, no. 1 (1928): 121–24; Commandant Ettori, "Le Berbère et le code forestier" (CHEAM: unpublished report, 1952); Jean-Joseph Souloumiac, "Le Problème du défrichement ou une politique forestière au Maroc" (CHEAM: unpublished report, 1947); and Jean-Joseph Souloumiac, "La Déforestation au Maroc" (CHEAM: unpublished report, 1948).

119. For details of the decree on grazing in the forests, see "Arrêté Viziriel du 15 Janvier 1921 Réglant le Mode d'Exercise du Droit au Parcours dans les Forêts Domaniales," Maroc/DAI/319a, CADN.

120. Rivet has criticized the way the protectorate authorities defined "Moroccan tribes" and "tribal boundaries." He explains that the "tribe" became

not only an administrative unit "but a museum of a network of alliances among men, fossilized by the imposition of an external and sovereign system of authority; therefore, in the end, a fiction." See Rivet, *Lyautey*, 2:202.

121. Limits on the numbers of livestock in a family herd are detailed in Boudy, *Économie forestière*, 1:587.

122. One economist has explained that the *tertib* was "no longer a mere tax, but also a political instrument for breaking down the native and historic method of utilizing the environment." See Melvin Knight, *Morocco as a French Economic Venture: A Study of Open Door Imperialism* (New York: D. Appleton-Century, 1937), 49. This was true of animal as well as plant agriculture in Morocco. As late as 1952, the government still was using this tactic. By raising the tertib on goats, the government sought to discourage goat production and increase sheep production, based on its fears that goats caused deforestation. See Maroc/DAI/#591, CADN.

123. Boudy, *Économie forestière*, 1:x.

124. Ibid., 1:627–28.

125. For examples of arguments for reducing migration, see Léon Lehureaux, "Comment s'effectue actuellement la transhumance," *L'Union Ovine* 2, no. 5 (1930): 189–91; Antonin Rolet, "La Transhumance en France: Ses difficultés," *L'Union Ovine* 2, no. 9 (1930): 386–89; Maurice Rondet-Saint, "Un Problème à résoudre en Afrique du nord: Le Nomadisme," *La Dépêche Coloniale*, 18 December 1931, 1. In fact, the example of errant, overgrazing herders in Algeria was frequently invoked during impassioned arguments that the French mountain peasants were environmentally destructive. See Tamara L. Whited, *Forests and Peasant Politics in Modern France* (New Haven, Conn.: Yale University Press, 2000), 203–5. Such arguments had been made in different forms since at least the 1880s. For one example, see Hippolyte Doumenjou, *Études sur la révision du code forestier: Les Reboisements en France et en Algérie* (Paris: J. Baudry, 1883). The French word *transhumance* generally means migration in a broader sense, not specifically moving herds up and down mountains for summer and winter pasturage.

126. Vétérinaire-Major Gadiou, "La Situation économique du Sous," *Revue de Géographie du Maroc* 6, no. 1 (1927): 137–65.

127. Commandant Ruet, "La Transhumance dans le moyen Atlas et la haute Moulouya" (CHEAM: unpublished report, 1952), 6.

128. Ibid.

129. For details on the creation and history of the department, see L.-A. Martin, "L'Élevage au Maroc," *Maroc Médical* 28, no. 292 (1949): 468–79.

130. Monod's personal information may be found in "Feuillet du Person-

nel," Dossier 8.752, Service Historique de l'Armée de la Terre, Vincennes, Paris (hereafter, SHAT). For more information on Monod and other French colonial veterinarians and their work, see Diana K. Davis, "Prescribing Progress: French Veterinary Medicine in the Service of Empire," *Veterinary Heritage* 29, no. 1 (2006): 1–7.

131. For details on the 1928 alfa grass reclassification, see Boudy, *Économie forestière*, 1:288.

132. Vétérinaire Colonel Théophile Monod, *L'Élevage au Maroc: Cours préparatoire au service des affaires indigènes* (Rabat: Résidence Générale de France au Maroc, Direction Générale des Affaires Indigènes, 1927), 10.

133. Descriptions of the "improvidence of the natives" may be found ibid. as well as in Service de l'Élevage du Maroc, "Historique du service de l'élevage du Maroc: Ses attributions et son rôle dans l'économie du pays," *Maroc Médical* 28, no. 292 (1949): 471–74; and Théophile Monod, "L'Élevage au Maroc," in *La Renaissance du Maroc: Dix ans de protectorat*, ed. Résidence Générale de la République Française au Maroc (Rabat: Résidence Générale de la République Française au Maroc, 1924), 294–306.

134. Capitaine Salvy, "La Crise du nomadisme dans le sud marocain" (CHEAM: unpublished report, 1949), 2.

135. Capitaine P. Azam, "La Structure politique et sociale de l'oued Dra" (CHEAM: unpublished report, 1947), 7. Azam was careful to explain that descendants of the original invading Arab nomads of centuries ago "remained nomads and gravitated around the Draa; they remain today like flies around a cake." Ibid., 8.

136. Service de l'Élevage du Maroc, "Historique du service," 472, provides details of the techniques of pasture reconstitution.

137. See Gudrun Dahl and Anders Hjort, *Pastoral Herd Growth and Household Economy*, vol. 2, Stockholm Series in Social Anthropology (Stockholm: Department of Social Anthropology, 1976), for a discussion of the goals of subsistence pastoralists. The traditional mode of extensive, mobile livestock production is well adapted to the arid North African/Mediterranean environment and is often the most efficient and environmentally friendly mode of resource use. See Michael B. Coughenour et al., "Energy Extraction and Use in a Nomadic Pastoral Ecosystem," *Science* 230, no. 4726 (1985): 619–25; Avi Perevolotsky and No'am Seligman, "Role of Grazing in Mediterranean Rangeland Ecosystems," *BioScience* 48, no. 12 (1998): 1007–17; Mark A. Blumler, "Successional Pattern and Landscape Sensitivity in the Mediterranean and Near East," in *Landscape Sensitivity*, ed. David S. G. Thomas and R. J Allison (Chichester: John Wiley, 1993), 287–305; and Ian Scoones, ed. *Living*

with Uncertainty: New Directions in Pastoral Development in Africa (London: Intermediate Technology Publications, 1995).

138. For relations with the military, see Monod, "L'Élevage au Maroc"; and Colonel Théophile Monod, "Le Rôle du vétérinaire colonial," *La Terre Marocaine* 1, no. 4 (1931): 4–15.

139. Boudy, *L'Arbre,* 40–41. For a discussion of the transfer of agricultural land to Europeans, see Swearingen, *Moroccan Mirages,* 143–44.

140. In Tunisia, this service is often called the conservation of water and soil service (Conservation des Eaux et Sols, CES). Bernard Heusch, "Cinquante ans de banquettes de DRS–CES en Afrique du nord: Un bilan," *Cahiers ORSTROM* 22, no. 2 (1986): 153–62. The DRS was inspired by the work of the American soil conservation service. See Henri Plateau, "La Lutte contre le ruissellement," *Bulletin de la Société des Sciences Naturelles du Maroc* 28, no. 1 (1948): 103. American experts like the former director of the soil and conservation service, W. C. Lowdermilk, conducted research consultations for the French colonial government and produced alarmist reports. See Lowdermilk's 1949 "Erosion et Conservation des Sols en Algérie," and his 1948 "L'Eau et la Conservation du Sol au Maroc." The latter is available in the library at the French National Forestry School at Nancy.

141. For details of the way the banquettes exacerbated erosion by acting like dams with no spillway during torrential rains, thus creating new gullies and even landslides, see Heusch, "Cinquante ans." This experience is sadly similar to others in Africa. See, for example, Kate B. Showers, *Imperial Gullies: Soil Erosion and Conservation in Lesotho* (Athens: Ohio University Press, 2005).

142. This narrative has also been very important in justifying legislation and stimulating the actual creation of numerous national parks in Algeria, Tunisia, and Morocco since the 1930s. For more information, see Diana K. Davis, "Environmentalism as Social Control? An Exploration of the Transformation of Pastoral Nomadic Societies in French Colonial North Africa," *Arab World Geographer* 3, no. 3 (2000): 182–98.

Chapter 6

1. Land statistics from Jean Saint-Germes, *Économie algérienne* (Alger: La Maison des Livres, 1955), 158. Population figures from John Ruedy, *Modern Algeria: The Origins and Development of a Nation* (Bloomington: Indiana University Press, 1992), 94.

2. Saint-Germes, *Économie,* 158.

3. Jean Saint-Germes, *La Réforme agraire algérienne* (Alger: La Maison des Livres, 1955), 5.

4. Saint-Germes, *Économie,* 157–58.

5. Paul Boudy, *Économie forestière nord-africaine: Description forestière de l'Algérie et de la Tunisie,* vol. 4 (Paris: Larose, 1955), 369.

6. M'hammed Boukhobza, *Monde rural: Contraintes et mutations* (Alger: Office des Publications Universitaires, 1992), 26.

7. Statistics from Charles-Robert Ageron, *Les Algériens musulmans et la France (1871–1919)* (Paris: Presses Universitaires de France, 1968), 2:806; and Djilali Sari, *La Dépossession des fellahs* (Alger: Société Nationale d'Édition et de Diffusion, 1978), 106.

8. Jean Poncet and André Raymond, *La Tunisie* (Paris: Presses Universitaires de France, 1971), 33 and 46.

9. Ibid., 33–34.

10. Boudy, *Économie forestière,* 4:461.

11. Poncet and Raymond, *La Tunisie,* 42.

12. Jean Brignon et al., *Histoire du Maroc* (Paris: Hatier, 1967), 378 and 380.

13. Will D. Swearingen, *Moroccan Mirages: Agrarian Dreams and Deceptions, 1912–1986* (Princeton, N.J.: Princeton University Press, 1987), 144.

14. Ibid., 141.

15. Paul Boudy, *Économie forestière nord-africaine: Milieu physique et milieu humain,* vol. 1 (Paris: Larose, 1948), 266–90.

16. Brignon et al., *Histoire,* 372–73.

17. Mohammad Riad El Ghonemy, *Land, Food, and Rural Development in North Africa.* Westview Special Studies in Social, Political and Economic Development (Boulder: Westview, 1993), 32. Population statistics for nomads are notoriously scarce in Morocco, for both the colonial and postcolonial periods. Nomads were not an officially recognized category in census counts during the twentieth century.

18. Although Algeria did not win its independence until 1962, a vicious war was waged between 1954 and 1962 that prevented any reliable statistics being gathered from 1954 until the early independence period.

19. Brignon et al., *Histoire,* 371.

20. Poncet and Raymond, *La Tunisie,* 47. Poncet does not include the wealthy, elite Tunisians in his estimate of average per capita income, which may account for why this figure is significantly lower than the similar figures for Morocco and Algeria.

21. Ibid., 47.

22. Mahfoud Bennoune, *The Making of Contemporary Algeria, 1830–1987: Colonial Upheavals and Post-independence Development* (Cambridge: Cambridge University Press, 1988), 90.

23. Ruedy, *Modern*, 123.

24. For a detailed overview of these processes, see Will D. Swearingen, "Is Drought Increasing in Northwest Africa? A Historical Analysis," in *The North African Environment at Risk*, ed. Will D. Swearingen and Abdellatif Bencherifa (Boulder, Colo.: Westview, 1996), 17–34. This author explains that the spread of cereal cultivation onto marginal lands was particularly widespread in the Maghreb from 1915 to 1928 because the French needed to produce more grain for France.

25. Wolfgang Trautmann, "The Nomads of Algeria under French Rule: A Study of Social and Economic Change," *Journal of Historical Geography* 15, no. 2 (1989): 126–38.

26. Boudy, *Économie forestière*, 4:559. These estimates, however, should be treated with caution since Boudy deduced the extent of earlier forests by using potential vegetation maps, and did not take regrowth into account in his calculations. Boudy assumed that Algeria had five million hectares of forest in 1830. The official statistics from 1872, however, show only 2,084,379 hectares of forested land at the time, and it is highly unlikely that three million hectares of forest had been lost during the first forty years of occupation. See Louis-François Victor Tassy, *Service forestier de l'Algérie, rapport adressé à m. le gouverneur de l'Algérie* (Paris: A. Hennuyer, 1872), 5. Despite the lack of accurate forestry statistics for North Africa until near the end of the nineteenth century, many scholars use Boudy's statistics, assuming they are based on authoritative evidence. The postcolonial scholar Djilali Sari, for instance, calculates that about two million hectares of forests were destroyed during the colonial period based on his reading of Boudy. See Sari, *La Dépossession*, 120.

27. Bennoune, *Making*, 89–90.

28. See, for instance, Service Psychologique, Xe Région Militaire, *Connaissance de l'Algérie* (Publications de l'Armée de la France, 1956). Available in the library at CAOM.

29. For example, Emberger's map of bioclimatic stages and phytogeographic map are both included as definitive sources for the presentation and analysis of Moroccan vegetation in the *Grand Encyclopedia of Morocco*, a compendium of "state of the art" knowledge of Morocco. See Mohamed Fennane, ed., *La Grande encyclopédie du Maroc: Flore et végétation* (Rabat: GEI, 1987), 9–11.

Other references are made to Emberger's works throughout this volume. Boudy's *Économie Forestière* also figures prominently in the encyclopedia's discussion of forests and forestry in Morocco. See ibid., 148–76. See also Ahmed Lahlimi Alami, ed., *La Grande encyclopédie du Maroc: Agriculture et pêche* (Rabat: GEI, 1987). Traces of the colonial environmental narrative may be found in many places in these two volumes.

30. For details of some of these projects, see Diana K. Davis, "Indigenous Knowledge and the Desertification Debate: Problematising Expert Knowledge in North Africa," *Geoforum* 36, no. 4 (2005): 509–24; and Slimane Bedrani, "Going Slow with Pastoral Cooperatives," *Ceres* 16, no. 4 (1983): 16–21.

31. Bedrani, "Going Slow."

32. Jean-Louis Ballais, "Aeolian Activity, Desertification and the 'Green Dam' in the Ziban Range, Algeria," in *Environmental Change in Drylands: Biogeographical and Geomorphological Perspectives,* ed. Andrew C. Millington and Ken Pye (New York: John Wiley, 1994), 177–98.

33. For a widely cited example in French, see Xavier de Planhol, *Les Fondements géographiques de l'histoire de l'Islam* (Paris: Flammarion, 1968), esp. part two of chap. 3, "La Ruine de la Maghreb."

34. J. V. Thirgood, *Man and the Mediterranean Forest: A History of Resource Depletion* (New York: Academic Press, 1981), 71. It should be noted that the other major work in English that has also been widely cited, John R. McNeill, *The Mountains of the Mediterranean World: An Environmental History* (New York: Cambridge University Press, 1992), mentions deforestation in Northern Morocco in a much more nuanced way that recognizes many of the complexities of and some of the arguments against the "myth of the Arab invasion." McNeill does, however, rely on the work of Boudy, Emberger, Maire, and Mikesell in his discussion of the potential vegetation and its deduction using relict vegetation in sacred groves, and thus does perpetuate part of the colonial narrative. Mikesell was one of the earliest postcolonial English-language scholars to incorporate much of the colonial declensionist narrative into his arguments. His influential article was published in the highly respected and widely read journal *Science.* See Marvin Mikesell, "Deforestation in Northern Morocco," *Science* 132, no. 3425 (1960): 441–48.

35. Salah E. Zaimeche, "The Consequences of Rapid Deforestation: A North African Example," *Ambio* 23, no. 2 (1994): 137.

36. Ibid., 136.

37. See, for example, Florian Plit, "La Dégradation de la végétation, l'érosion et la lutte pour protéger le milieu naturel en Algérie et au Maroc," *Méditerranée* 49, no. 3 (1983): 79–88; Armand Pons and Pierre Quezel, "The History

of the Flora and Vegetation and Past and Present Human Disturbance in the Mediterranean Region," in *Plant Conservation in the Mediterranean Area,* ed. César Gomez-Campo (Dordrecht: Dr. W. Junk, 1985), 25–43; M. Barbero et al., "Changes and Disturbances of Forest Ecosystems Caused by Human Activities in the Western Part of the Mediterranean Basin," *Vegetatio* 87, no. 2 (1990): 151–73; Abdelmalik Benabid, *Flore et écosystèmes du Maroc: Évaluation et préservation de la biodiversité* (Paris: Ibis, 2000); Daniel Vaxelaire, ed., *La Grande encyclopédie du Maroc* (Rabat: GEI, 1987); Oreste Reale and Paul Dirmeyer, "Modeling Effects of Vegetation on Mediterranean Climate during the Roman Classical Period," *Global and Planetary Change* 25, no. 3–4 (2000): 163–84; Fouad Zaim, "Déforestation et érosion dans le Maroc méditérranéen, effets socio-économiques," in *La Forêt marocaine: Droit, économie, écologie,* ed. SOMADE (Casablanca: Afrique Orient, 1989), 95–109; République Algérienne Democratique et Populaire, Direction des Forêts, *Rapport national de l'Algérie sur la mise en oeuvre de la convention de lutte contre la désertification* (Alger: Ministère de l'Agriculture et du Developpement Rural, Direction des Forêts, 2004).

38. For examples and discussion, see David S. G. Thomas and Nicholas Middleton, *Desertification: Exploding the Myth* (West Sussex: John Wiley, 1994); and Jeremy Swift, "Desertification: Narratives, Winners and Losers," in *The Lie of the Land: Challenging Received Wisdom on the African Environment,* ed. Melissa Leach and Robin Mearns (London: International African Institute, 1996), 73–90.

39. See, for instance, James Fairhead and Melissa Leach, *Reframing Deforestation: Global Analyses and Local Realities; Studies in West Africa* (London: Routledge, 1998), 188; Swift, "Desertification"; David S. G. Thomas, "Science and the Desertification Debate," *Journal of Arid Environments* 37, no. 4 (1997): 599–608; Andrew Goudie, *The Human Impact on the Natural Environment* (Cambridge, Mass.: MIT Press, 2000), 68; and Graeme Barker, "A Tale of Two Deserts: Contrasting Desertification Histories on Rome's Desert Frontiers," *World Archaeology* 33, no. 3 (2002): 488–507.

40. Louis Lavauden, "Les Forêts du Sahara," *Revue des Eaux et Forêts* 65, no. 6 (1927): 267, 337. To the best of my knowledge, the eminent arid lands ecologist Henry Le Houérou was the first to publish on the use of the word "desertification" by Lavauden. See Henry N. Le Houérou, "Man-Made Deserts: Desertization Processes and Threats," *Arid Land Research and Management* 16, no. 1 (2002): 1–36. I am very grateful to Henry for sharing his knowledge of desertification in North Africa with me.

41. J. Reynard, *Restauration des forêts et des pâturages du sud de l'Algérie (province d'Alger)* (Alger: Adolphe Jourdan, 1880), 15.

42. Jean Despois, *La Tunisie* (Paris: Larousse, 1930), 34. A few pages later (55) Despois claims that this desertification commenced with the Hillalian Arab "invasion" of the eleventh century and quotes Ibn Khaldoun to describe their migration as a "plague of locusts." Many books such as this were published in the 1930s that emphasized the declensionist environmental narrative and frequently cited Ibn Khaldoun. One of the most influential was the geographer Émile Gautier's book on the history of North Africa, *Le Passé de l'Afrique du nord: Les Siècles obscurs* (Paris: Payot, 1942). Augustin Bernard, in his volume on North Africa in the widely read *Universal Geography*, also repeated the colonial narrative and quoted Ibn Khaldoun to argue that "every country conquered by the Arabs is a ruined country." See Augustin Bernard, *Afrique septentrionale et occidentale. Première partie: Généralités—Afrique du nord*, vol. 1 of *Géographie universelle*, ed. Paul Vidal de la Blache (Paris: Armand Colin, 1937), 65, 75–79.

43. Capitaine Salvy, "La Crise du nomadisme dans le sud marocain" (CHEAM: unpublished report, 1949), 10.

44. In Morocco, for example, the colonial declensionist narrative is currently being used to implement along neoliberal lines an agricultural restructuring that is predicted to impoverish a large segment of Morocco's cereal farmers. It will likely have the same effect on pastoral nomads and others who depend on the remaining collective lands. See Diana K. Davis, "Neoliberalism, Environmentalism and Agricultural Restructuring in Morocco," *Geographical Journal* 172, no. 2 (2006): 88–105.

45. Anon., "Land Degradation (Threats of Desertification)," *UN Chronicle* 34, no. 2 (1997): 26–28.

46. For an excellent analysis of the problems with global contemporary desertification narratives, the antidesertification industry, and the questionable uses to which such narratives are often put, see William M. Adams, *Green Development: Environment and Sustainability in the Third World* (London: Routledge, 2001), 175–214; and Swift, "Desertification."

47. For a recent analysis of desertification claims regarding Morocco and evidence that undermines them, see Davis, "Indigenous Knowledge." For an example of the legacy of the declensionist narrative in Algeria, see République Algérienne Democratique et Populaire, Direction des Forêts, *Rapport National*. Similar reports are available for all three Maghreb countries. Much of the money currently devoted to these antidesertification projects could be

much better spent on economic and social development, especially animal health, which is woefully underfunded and neglected in many of the pastoral regions of the Maghreb.

48. John R. McNeill, "Observations on the Nature and Culture of Environmental History," *History and Theory* 42, no. 4 (December 2003): 5–43, esp. 25–28 and 30.

49. William C. Brice, ed., *Environmental History of the Near and Middle East since the Last Ice Age* (London: Academic Press, 1978); J. Malcolm Wagstaff, *The Evolution of Middle Eastern Landscapes: An Outline to A.D. 1840* (London: Croom Helm, 1985): Peter Christiansen, *The Decline of Iranshahr: Irrigation and Environments in the History of the Middle East 500 B.C. to A.D. 1500* (Copenhagen: Museum Tusculanum, 1993). Shaul Cohen's book on the importance of planting for the control of land in Jerusalem might be added here, although it is less an environmental history than a deft political analysis. See Shaul E. Cohen, *The Politics of Planting: Jewish-Palestinian Competition for Control of Land in the Jerusalem Periphery* (Chicago: University of Chicago Press, 1993).

50. See, for some examples, James Fairhead and Melissa Leach, *Misreading the African Landscape: Society and Ecology in a Forest-Savanna Mosaic* (Cambridge: Cambridge University Press, 1996); Fairhead and Leach, *Reframing Deforestation;* Melissa Leach and Robin Mearns, eds. *The Lie of the Land: Challenging Received Wisdom on the African Environment* (London: International African Institute, 1996); David M. Anderson, *Eroding the Commons: The Politics of Ecology in Baringo, Kenya, 1890–1963* (Oxford: James Currey, 2002); Thomas J. Bassett and Donald Crummey, eds. *African Savannas: Global Narratives and Local Knowledge of Environmental Change* (Oxford: James Currey, 2003); Richard H. Grove, *Green Imperialism: Colonial Expansion, Tropical Island Edens, and the Origins of Environmentalism, 1600–1860* (Cambridge: Cambridge University Press, 1995); Christian Kull, "Deforestation, Erosion, and Fire: Degradation Myths in the Environmental History of Madagascar," *Environment and History* 6, no. 4 (2000): 423–50; James McCann, *Green Land, Brown Land, Black Land: An Environmental History of Africa* (Portsmouth, N.H.: Heinemann, 1999); Roderick P. Neumann, *Imposing Wilderness: Struggles over Livelihood and Nature Preservation in Africa* (Berkeley: University of California Press, 1998); Nancy L. Peluso, *Rich Forests, Poor People: Resource Control and Resistance in Java* (Berkeley: University of California Press, 1992); Mary Tiffen, Michael Mortimore, and Francis Gichuki, *More People, Less Erosion: Environmental Recovery in Kenya* (Chichester: John Wiley, 1994); Kate B. Showers, *Imperial Gullies: Soil Erosion and Conservation*

in Lesotho (Athens: Ohio University Press, 2005); James L. A. Webb, *Tropical Pioneers: Human Agency and Ecological Change in the Highlands of Sri Lanka, 1800–1900* (Athens: Ohio University Press, 2002); Ramachandra Guha, *The Unquiet Woods: Ecological Change and Peasant Resistance in the Himalaya* (New Delhi: Oxford University Press, 1989); Richard H. Grove, Vinita Damodaran, and Satpal Sangwan, eds., *Nature and the Orient: The Environmental History of South and Southeast Asia* (Delhi: Oxford University Press, 1998); and S. Ravi Rajan, *Modernizing Nature: Forestry and Imperial Eco-Development 1800–1950* (Oxford: Clarendon, 2006).

51. Auguste Chevalier, *L'Agronomie coloniale et le Muséum National d'Histoire Naturelle: Premières conférences du cours sur les productions coloniales végétales et l'agronomie tropicale* (Paris: Laboratoire d'Agronomie Coloniale, 1930), 2.

52. For one example, see André Jacquot, *Incendies en forêt (Forest Fires)*, trans. C. E. C. Fischer (Calcutta: Superintendent Government Printing, 1910), available at the India Office of the British Library, V/27/560/12.

53. See McNeill, "Observations," 29; and Caroline Ford, "Landscape and Environment in French Historical and Geographical Thought: New Directions," *French Historical Studies* 24, no. 1 (2001): 125–34.

54. See, for example, Ageron, *Les Algériens;* Charles-André Julien, *Histoire de l'Algérie contemporaine: La Conquête et les débuts de la colonisation (1827–1871)* (Paris: Presses Universitaires de France, 1964); Yves Lacoste, André Nouschi, and André Prenant, *L'Algérie: Passé et présent* (Paris: Éditions Sociales, 1960); Sari, *La Dépossession;* Boukhobza, *Monde rural;* Daniel Rivet, *Lyautey et l'instauration du protectorat français au Maroc 1912–1925,* 3 vols. (Paris: Harmattan, 1988); Brignon et al., *Histoire;* and Poncet and Raymond, *La Tunisie,* among many others. Nearly all of these authors recognized the profound importance of the colonial forest laws, but none of them analyzed the environmental narratives that informed and influenced these laws and how they were implemented.

55. A recent UNFAO report on forest resources in the Mediterranean region cites Boudy's work and includes calculations of "potential and actual forest areas." See Food and Agriculture Organization of the United Nations, *Forest Resources Assessment 1990: Non-Tropical Developing Countries Mediterranean Region* (Rome: FAO, 1994), esp. table 4. The statistics for the citations of Boudy come from a search solely for his four-volume *Économie Forestière* conducted 15 August 2005 on the database "Science Citation Index." Philibert Guinier, the "founder" of forest ecology in France, considered Boudy's work so important that he wrote that it was valid not only for the

greater Mediterranean basin but also for "all regions with an arid climate." See Philibert Guinier, "Nécrologie: Paul Boudy (1874–1957)," *Revue Forestière Française* 3 (May 1958): 221.

56. A search for citations of Emberger's publications on the "Science Citation Index" returned 482 citations between the years 1975 and 2005 (seven of these were from 2005). The search was conducted 16 August 2005.

57. For more details, see United Nations Educational, Scientific, and Cultural Organization, *Bioclimatic Map of the Mediterranean Zone. Explanatory Notes,* vol. 21 of *Ecological Study of the Mediterranean Zone* (Paris: UNESCO–FAO, 1963).

58. For more details, see United Nations Educational, Scientific, and Cultural Organization, *Vegetation Map of the Mediterranean Zone. Explanatory Notes,* vol. 30 of *Ecological Study of the Mediterranean Zone* (Paris: UNESCO–FAO, 1969).

59. See Frank White, *The Vegetation of Africa: A Descriptive Memoir to Accompany the UNESCO/AETFAT/UNSO Vegetation Map of Africa* (Paris: UNESCO, 1983), esp. 225–31. The work of Emberger, Boudy's volumes on forestry, and research by others trained in the French school of phytoecology supplied the bulk of information for the section on North Africa.

60. See Louis Emberger, "Aperçu général [de la végétation]," in *La Science au Maroc,* ed. Paul Boudy and P. Despujols (Casablanca: Imprimeries Réunies, 1934), 152; and I. Nahal, "The Mediterranean Climate from a Biological Viewpoint," in *Mediterranean-Type Shrublands,* ed. Francesco di Castri, David W. Goodall, and Raymond Specht (New York: Elsevier Scientific, 1981), 66–71.

61. Emberger's influence was also substantially increased by the large number of advanced students he directed at the university in Montpellier over the course of his career. A dozen or more of these students became significant researchers and scholars and they have published many books and articles that have carried parts of the declensionist narrative forward to the present and generalized it to Mediterranean ecosystems. Many of these former students have published on the ecology of the Mediterranean basin and most of them relied on and used Emberger's work on Mediterranean ecology, potential vegetation, and bioclimatic maps. The volume on Mediterranean-type shrublands in what was for over a decade the definitive series on the ecosystems of the world, for example, contained nine chapters (one-third of the book) by Emberger's former students. See Francesco Di Castri, David W. Goodall, and Raymond Specht, eds., *Mediterranean-Type Shrublands* (New York: Elsevier Scientific, 1981), chaps. 3–8, 19, 25, and 26.

62. Fairhead and Leach, *Reframing Deforestation,* 164–70. See also Kull, "Deforestation," 429. But see Kate B. Showers, "Gallery: Kate B. Showers on Mapping African Soils," *Environmental History* 10, no. 2 (2005): 314–20, for a powerful and elegant deconstruction of the use of African soil maps to provide scientific authority, similar to that provided by potential-vegetation maps.

63. For excellent discussions of many of the problems associated with these new environmental technologies, see Paul Robbins, "Fixed Categories in a Portable Landscape: The Causes and Consequences of Land-Cover Categorization," *Environment and Planning A* 33, no. 1 (2001): 161–79; and Matthew D. Turner, "Methodological Reflections on the Use of Remote Sensing and Geographic Information Science in Human Ecological Research," *Human Ecology* 31, no. 2 (2003): 255–79.

64. Douglas R. Weiner, "A Death-Defying Attempt to Articulate a Coherent Definition of Environmental History," *Environmental History* 10, no. 3 (2005): 417.

65. Other examples are available from other parts of the Mediterranean basin. In Israel, for instance, the application of principles similar to those followed by the colonial and postcolonial administrations in North Africa has resulted in the creation of near monostands of thick woods that "[exclude] many local species of native fauna and flora, leading environmentalists to label such thickets 'green deserts.'" See Avi Perevolotsky and No'am Seligman, "Role of Grazing in Mediterranean Rangeland Ecosystems," *BioScience* 48, no. 12 (1998): 1011. In southern France, not only has the suppression of grazing caused herders to lose their livelihoods, but it has also resulted in significant problems with wildfires in the region. See Eileen O'Rourke, "Changing Identities, Changing Landscapes: Human-Land Relations in Transition in the Aspre, Roussillon," *Ecumene* 6, no. 1 (1999): 29–50.

Appendix

1. The primary sources for this overview are Gerald Blake, John Dewdney, and Jonathan Mitchell, *The Cambridge Atlas of the Middle East and North Africa* (Cambridge: Cambridge University Press, 1987); Jean Despois and René Raynal, *Géographie de l'Afrique du nord-ouest* (Paris: Payot, 1967); Harold Nelson, *Morocco: A Country Study* (Washington, D.C.: American University, 1985); Harold Nelson, *Algeria: A Country Study* (Washington, D.C.: American University, 1985); Harold Nelson, *Tunisia: A Country Study* (Washington, D.C.: American University, 1988); United Kingdom, Naval Intelligence Division,

Algeria, Geographical Handbook Series (Oxford: University Press, 1943); United Kingdom, Naval Intelligence Division, *Tunisia,* Geographical Handbook Series (Oxford: University Press, 1945): and United Kingdom, Naval Intelligence Division, *Morocco, Geographical Handbook Series* (Oxford: University Press, 1941). Current statistics for agricultural areas and land use come from the searchable data set at the UN Food and Agricultural Organization, FAO-STAT, available at http://faostat.fao.org.

2. Anon., *Statistique et documents relatifs au Sénatus-Consulte sur la propriété arabe* (Paris: Imprimerie Impériale, 1863), 478. This figure does not include land lying fallow at that time. In the early colonial and precolonial period, for which few reliable estimates of agricultural production are available, land use tended to vacillate between farming and herding depending on both weather conditions and the international market for agricultural products. Due to local ecological conditions, land considered arable can change significantly in extent from year to year throughout the Maghreb.

3. Jean Saint-Germes, *Économie algérienne* (Alger: La Maison des Livres, 1955), 95. Although the production of crops other than cereals, especially vines, had grown significantly during the colonial period, in the 1950s cereals still accounted for at least 75 percent of land planted to crops.

4. Jean Saint-Germes, *La Réforme agraire algérienne* (Alger: La Maison des Livres, 1955), 5.

5. Jean Poncet and André Raymond, *La Tunisie* (Paris: Presses Universitaires de France, 1971), 39–40.

6. Will D. Swearingen, *Moroccan Mirages: Agrarian Dreams and Deceptions, 1912–1986* (Princeton, N.J.: Princeton University Press, 1987), 22, 147.

7. For a detailed account of the development of the agricultural sector in Morocco, including the highly productive irrigated sector and dam construction, see ibid.

8. Bernard Dell, Angus J. Hopkins, and Byron B. Lamont, eds. *Resilience in Mediterranean-Type Ecosystems* (Dordrecht: Dr. W. Junk, 1986). See also Yitzchak Gutterman, *Regeneration of Plants in Arid Ecosystems Resulting from Patch Disturbance* (Dordrecht: Kluwer Academic, 2001).

9. For information on adaptations to aridity and drought, see Mark A. Blumler, "Biogeography of Land-Use Impacts in the Near East," in *Nature's Geography: New Lessons for Conservation in Developing Countries,* ed. Karl Zimmerer and Kenneth Young (Madison: University of Wisconsin Press, 1998), 215–36; Kamal A. Batanouny, *Plants in the Deserts of the Middle East* (Berlin: Springer-Verlag, 2001); and Francesco Di Castri, David W. Goodall, and Ray-

mond Specht, eds., *Mediterranean-Type Shrublands* (Amsterdam: Elsevier Scientific, 1981).

10. Driss Ouaskioud, "Contribution à l'étude de la dynamique de la végétation steppique après une mise en défens de longue durée: Cas de la station d'amélioration pastorale Anbad Boumalne Dades (Ouarzazate)," Mémoire de Troisième Cycle, Institut Agronomique et Vétérinaire Hassan II, 1999; and D. A. Thomas et al., "Rangeland Regeneration in the Steppic Regions of the Mediterranean Basin," in *Rangelands: A Resource under Siege,* ed. P. L. Joss (Cambridge: Cambridge University Press, 1986), 280–87.

11. M. E. Figueroa and A. J. Davy, "Response of Mediterranean Grassland Species to Changing Rainfall," *Journal of Ecology* 79, no. 4 (1991): 938. See also Peter Haase, Francisco Pugnaire, and L. D. Incoll, "Seed Production and Dispersal in the Semi-arid Tussock Grass *Stipa tenacissima* L. during Masting," *Journal of Arid Environments* 31, no. 1 (1995): 55–65.

12. For details on fire adaptations, see Michael G. Barbour et al., *Terrestrial Plant Ecology,* 3rd ed. (Menlo Park, Calif.: Addison, Wesley, Longman, 1999), 459–64; and Louis Trabaud, "Postfire Plant Community Dynamics in the Mediterranean Basin," in *The Role of Fire in Mediterranean-Type Ecosystems,* ed. José Moreno and Walter Oechel (New York: Springer-Verlag, 1994), 1–15.

13. A. Schoenenberger, *Utilisation de la végétation naturelle dans la vallée du Dra: Lutte contre la désertification par une amélioration sylvo-pastorale* (Zagora: PROLUDRA/ORMVA, 1994). Indeed, it has been demonstrated that some plants will actually die if they are not grazed for a long period of time. See Gufu Oba, Nils Stenseth, and Walter Lusigi, "New Perspectives on Sustainable Grazing Management in Arid Zones of Sub-Saharan Africa," *BioScience* 50, no. 1 (2000): 35–51.

14. S. M. Bouzid and V. P. Papanastis, "Effects of Seeding Rate and Fertilizer on Establishment and Growth of *Atriplex halimus* and *Medicago arborea,*" *Journal of Arid Environments* 33, no. 1 (1996): 109–15.

15. C. N. Tsiouvaras, B. Noitsakis, and V. P. Papanastasis, "Clipping Intensity Improves Growth Rate of Kermes Oak Twigs," *Forest Ecology and Management* 15, no. 3 (1986): 229–37.

16. Michael B. Coughenour, "Graminoid Responses to Grazing by Large Herbivores: Adaptations, Exaptations, and Interacting Processes," *Annals of the Missouri Botanical Garden* 72, no. 4 (1985): 852–63; and Richard Mack and John Thompson, "Evolution in Steppe with a Few Large, Hooved Mammals," *American Naturalist* 119, no. 6 (1982): 757–73.

17. Figueroa, "Response"; and Batanouny, *Plants in the Deserts.*

18. Douglas G. Sprugel, "Disturbance, Equilibrium, and Environmental Variability: What is 'Natural' Vegetation in a Changing Environment?" *Biological Conservation* 58, no. 1 (1991): 1–18.

19. Daniel Botkin, *Discordant Harmonies: A New Ecology for the Twenty-first Century* (New York: Oxford University Press, 1990); Alan Hastings et al., "Chaos in Ecology: Is Mother Nature a Strange Attractor?" *Annual Review of Ecology and Systematics* 24 (1993): 1–33; and Robert McIntosh, "Pluralism in Ecology," *Annual Review of Ecology and Systematics* 18 (1987): 321–41.

20. See Blumler, "Biogeography," for a good overview of these topics.

21. Walter Goldschmidt, "The Failure of Pastoral Economic Development Projects in Africa," in *The Future of Pastoral Peoples,* ed. John Galaty (Ottawa: International Development Research Center, 1981), 101–18; Maryam Niamir-Fuller, "The Resilience of Pastoral Herding in Sahelian Africa," in *Linking Social and Ecological Systems: Management Practices and Social Mechanisms,* ed. Fikret Berkes and Carl Folke (Cambridge: Cambridge University Press, 1998), 250–84; Ian Scoones, "Range Management Science and Policy: Politics, Polemics and Pasture in Southern Africa," in *The Lie of the Land: Challenging Received Wisdom on the African Environment,* ed. Melissa Leach and Robin Mearns (London: International African Institute, 1996), 34–53; and Andrew Warren, "Changing Understandings of African Pastoralism and the Nature of Environmental Paradigms," *Transactions of the Institute of British Geographers* 20, no. 2 (1995): 193–203.

22. Jerrold L. Dodd, "Desertification and Degradation in Sub-Saharan Africa: The Role of Livestock," *BioScience* 44, no. 1 (1994): 28–34; James E. Ellis and David M. Swift, "Stability of African Pastoral Ecosystems: Alternate Paradigms and Implications for Development," *Journal of Range Management* 41, no. 6 (1988): 450–59; and Mark Westoby, Brian Walker, and Immanuel Noy-Meir, "Opportunistic Management of Rangelands Not at Equilibrium," *Journal of Range Management* 42, no. 4 (1989): 266–74.

23. Blumler, "Biogeography."

24. Other recent research has shown that in North Africa and the Middle East vegetation growth is significant in areas that are fenced off for only a few years, indicating that these environments are not "overgrazed and degraded" but, more likely, simply being heavily but sustainably grazed. See Henry N. Le Houérou, "Restoration and Rehabilitation of Arid and Semiarid Mediterranean Ecosystems in North Africa and West Asia: A Review," *Arid Soil Research and Rehabilitation* 14, no. 1 (2000): 3–14. A global overview that includes North Africa concludes that the effects of grazing on vegetation and soils are significantly less damaging than generally claimed. See Daniel G.

Milchunas and William K. Lauenroth, "Quantitative Effects of Grazing on Vegetation and Soils over a Global Range of Environments," *Ecological Applications* 63, no. 4 (1993): 327–66.

25. Niamir-Fuller, "Resilience"; Diana K. Davis, "Indigenous Knowledge and the Desertification Debate: Problematising Expert Knowledge in North Africa," *Geoforum* 36, no. 4 (2005): 509–24; Hsain Ilahiane, "The Berber Agdal Institution: Indigenous Range Management in the Atlas Mountains," *Ethnology* 38, no. 1 (1999): 21–46; Matthew D. Turner, "Overstocking the Range: A Critical Analysis of the Environmental Science of Sahelian Pastoralism," *Economic Geography* 69, no. 4 (1993): 402–21; and Jeremy Swift, "Dynamic Ecological Systems and the Administration of Pastoral Development," in *Living with Uncertainty: New Directions for Pastoral Development in Africa,* ed. Ian Scoones (London: Intermediate Technology Publications, 1995), 153–73; and Linda Olsvig-Whittaker, Eliezer Frankenberg, Avi Perevolotsky, and Eugene D. Ungar, "Grazing, Overgrazing, and Conservation: Changing Concepts and Practices in the Negev Rangelands," *Sécheresse* 17, no. 1–2 (2006): 195–99.

26. Westoby, "Opportunistic Management," 197.

Bibliography

Abun-Nasr, Jamil M. *A History of the Maghrib.* Cambridge: Cambridge University Press, 1971.

Adams, William M. *Green Development: Environment and Sustainability in the Third World.* London: Routledge, 2001.

Ageron, Charles-Robert. *Les Algériens musulmans et la France (1871–1919).* 2 vols. Paris: Presses Universitaires de France, 1968.

———. *Histoire de l'Algérie contemporaine: De l'insurrection de 1871 au déclenchement de la guerre de libération (1954).* Paris: Presses Universitaires de France, 1979.

———. *Modern Algeria: A History from 1830 to the Present.* Translated by Michael Brett. London: Hearst, 1991.

AIGREF. See Association des Ingénieurs du Génie Rural, des Eaux et des Forêts.

Aldrich, Robert. *Greater France: A History of French Overseas Expansion.* London: Macmillan, 1996.

Alzonne, Clément. *L'Algérie.* Paris: Fernand Nathan, 1937.

Anderson, David M. *Eroding the Commons: The Politics of Ecology in Baringo, Kenya, 1890–1963.* Oxford: James Currey, 2002.

Anon. "La Forêt marocaine." In *Études marocaines: Supplément au bulletin d'informations et de documentation de la résidence générale.* Rabat: Résidence Générale de France au Maroc, 1940, 5–28.

———. "Land Degradation (Threats of Desertification)." *UN Chronicle* 34, no. 2 (1997): 26–28.

———. *Statistique et documents relatifs au Sénatus-Consulte sur la propriété arabe.* Paris: Imprimerie Impériale, 1863.

———. *La Vie studieuse, féconde et désintéressée de Paul Bourde, publiciste colonial, agronome, rénovateur des oliveraies tunisiennes.* Paris: Montsouris, 1943.

Association des Ingénieurs du Génie Rural, des Eaux et des Forêts. *Des Officiers royaux aux ingénieurs d'état dans la France rurale (1219–1965)*. Paris: Éditions Technique et Documentation, 2001.

Association Française pour l'Avancement des Sciences. *Compte rendu de la 10e session, Alger 1881*. Paris: Secrétariat de l'Association, 1882.

Attia, Habib. "L'Évolution des structures sociales et économiques dans les hautes steppes." *Revue Tunisienne des Sciences Sociales* 3, no. 6 (1966): 5–38.

d'Avezac, Marie Armand. "Esquisse générale de l'Afrique et Afrique ancienne." In *L'Afrique*, 1–48. Paris: Firmin Didot Frères, 1844.

Azam, P. "La Structure politique et sociale de l'oued Dra." CHEAM: unpublished report, 1947.

Ballais, Jean-Louis. "Aeolian Activity, Desertification and the 'Green Dam' in the Ziban Range, Algeria." In *Environmental Change in Drylands: Biogeographical and Geomorphological Perspectives*, edited by Andrew C. Millington and Ken Pye, 177–98. New York: John Wiley, 1994.

———. "Conquests and Land Degradation in the Eastern Maghreb During Classical Antiquity and the Middle Ages." In *The Archaeology of Drylands*, edited by Graeme Barker and David Gilbertson, 125–136. London: Routledge, 2000.

Ballouche, Aziz, D. Lefevre, C. Carruesco, J. Raynal, and J. Texier. "Holocene Environments of Coastal and Continental Morocco." In *Quaternary Climate in Western Mediterranean*, edited by Fernando Lopez-Vera, 517–31. Madrid: Universidad Autónoma de Madrid, 1986.

Barbero, M., G. Bonin, R. Loisel, and P. Quézel. "Changes and Disturbances of Forest Ecosystems Caused by Human Activities in the Western Part of the Mediterranean Basin." *Vegetatio* 87, no. 2 (1990): 151–73.

Barbour, Michael G., Jack H. Burk, Wanna D. Pitts, Frank S. Gilliam, and Mark W. Schwartz. *Terrestrial Plant Ecology*. 3rd ed. Menlo Park, Calif.: Addison, Wesley, Longman, 1999.

Barker, Graeme. "A Tale of Two Deserts: Contrasting Desertification Histories on Rome's Desert Frontiers." *World Archaeology* 33, no. 3 (2002): 488–507.

Bassett, Thomas J., and Donald Crummey, eds. *African Savannas: Global Narratives and Local Knowledge of Environmental Change*. Oxford: James Currey, 2003.

Batanouny, Kamal A. *Plants in the Deserts of the Middle East*. Berlin: Springer-Verlag, 2001.

Battandier, Jules-Aimée, and Louis Trabut. *L'Algérie: Le Sol et les habitants, flore, faune, géologie, anthropologie, ressources agricoles et économiques*. Paris: J.-B. Baillière, 1898.

Battistini, Eugène. *Les Forêts de chêne-liège de l'Algérie*. Alger: Victor Heintz, 1937.

Becquerel, Antoine César. *Des Climats et de l'influence qu'exercent les sols boisés et non boisés*. Paris: Firmin Didot Frères, 1853.

Bedrani, Slimane. "Going Slow with Pastoral Cooperatives." *Ceres* 16, no. 4 (1983): 16–21.

Benabid, Abdelmalik. *Flore et écosystèmes du Maroc: Évaluation et préservation de la biodiversité*. Paris: Ibis, 2000.

Benjamin, Roger. *Orientalist Aesthetics: Art, Colonialism, and French North Africa, 1880–1930*. Berkeley: University of California Press, 2003.

Benjamin, Roger, ed. *Orientalism: Delacroix to Klee*. Sydney: Art Gallery of New South Wales, 1997.

Bennoune, Mahfoud. *The Making of Contemporary Algeria, 1830–1987: Colonial Upheavals and Post-independence Development*. Cambridge: Cambridge University Press, 1988.

Bergeret, Anne. "Discours et politique forestières coloniales en Afrique et à Madagascar." *Revue Française d'Histoire d'Outre-Mer* 79, no. 298 (1993): 23–47.

Bernard, Augustin. *Afrique septentrionale et occidentale. Première partie: Généralités—Afrique du nord*. Vol. 11 of *Géographie universelle*, edited by Paul Vidal de la Blache. Paris: Armand Colin, 1937.

———. *Une Mission au Maroc*. Paris: Le Comité du Maroc, 1904.

Bernard, Augustin, and Napoléon Lacroix. *L'Évolution du nomadisme en Algérie*. Alger: Adolphe Jourdan, 1906.

Bernard, J., and M. Reille. "Nouvelles analyses polliniques dans l'Atlas de Marrakech, Maroc." *Pollen et Spores* 29, no. 2–3 (1987): 225–40.

Bernis, Jean. *Quelques considérations sur la colonisation de l'Algérie*. Toulouse: Ph. Montaubin, 1866.

Bertherand, Eugène L. *L'Eucalyptus au point de vue de l'hygiène en Algérie*. Alger: Victor Aillaud, 1876.

Besnier, Maurice. *Géographie ancienne du Maroc*. Paris: Ernest Leroux, 1904.

———. "La Géographie économique du Maroc dans l'antiquité." *Archives Marocaines* 7, no. 1 (1906): 271–93.

Betts, Raymond. *Assimilation and Association in French Colonial Theory, 1890–1914*. New York: Columbia University Press, 1961.

Blake, Gerald, John Dewdney, and Jonathan Mitchell. *The Cambridge Atlas of the Middle East and North Africa*. Cambridge: Cambridge University Press, 1987.

Blottière, Jean. *Les Productions algériennes.* Cahiers du centenair de l'Algérie. Orléans: Imprimerie A Pigelet [1931].

Blumler, Mark A. "Biogeography of Land-Use Impacts in the Near East." In *Nature's Geography: New Lessons for Conservation in Developing Countries,* edited by Karl Zimmerer and Kenneth Young, 215–36. Madison: University of Wisconsin Press, 1998.

———. "Successional Pattern and Landscape Sensitivity in the Mediterranean and Near East." In *Landscape Sensitivity,* edited by David S. G. Thomas and R. J Allison, 287–305. Chichester: John Wiley, 1993.

Boissier, Gaston. *L'Afrique romaine: Promenades archéologiques en Algérie et en Tunisie.* Paris: Hachette, 1895.

Bosch, J. M., and J. D. Hewlett. "A Review of Catchment Experiments to Determine the Effects of Vegetation Changes on Water Yield and Evapotranspiration." *Journal of Hydrology* 55, no. 1–4 (1982): 3–23.

Botkin, Daniel. *Discordant Harmonies: A New Ecology for the Twenty-first Century.* New York: Oxford University Press, 1990.

Boudy, Paul. *L'Arbre et les forêts au Maroc. Cours préparatoire au service des affaires indigènes.* Rabat: Résidence Générale de France au Maroc, Direction Générale des Affaires Indigènes, 1927.

———. *Économie forestière nord-africaine: Description forestière de l'Algérie et de la Tunisie.* Vol. 4. Paris: Larose, 1955.

———. *Économie forestière nord-africaine: Description forestière du Maroc.* Vol. 3. 2nd ed. Paris: Larose, 1958.

———. *Économie forestière nord-africaine: Milieu physique et milieu humain.* Vol. 1. Paris: Larose, 1948.

———. "Les Forêts du Maroc." *La Revue Générale des Sciences Pures et Appliquées* 25, no. 7 (1914): 350–54.

———. "L'Oeuvre forestière française au Maroc." *Bulletin de la Société Forestière de Franche-Comté et des Provinces de l'Est* 27, no. 1 (1954): 1–10.

Boudy, Paul, and Louis Emberger. "La Forêt marocaine." In *La Science au Maroc,* edited by Paul Boudy and P. Despujols, 191–206. Casablanca: Imprimeries Réunies, 1934.

Boukhobza, M'hammed. *Monde rural: Contraintes et mutations.* Alger: Office des Publications Universitaires, 1992.

Bourde, Paul. *Rapport sur les cultures fruitières et en particulier sur la culture de l'olivier dans le centre de la Tunisie.* Tunis: Imprimerie Générale (Picard), 1899.

Boussingault, Jean Baptiste. *Économie rurale considérée dans ses rapports avec la chimie, la physique et la météorologie.* Vol. 2. 2nd ed. Paris: Béchet Jeune, 1851.

Boutilly, V. *Recueil de la législation forestière algérienne, lois, décrets et règlements divers.* Paris: Berger-Levrault, 1905.

Bouzid, S. M., and V. P. Papanastis. "Effects of Seeding Rate and Fertilizer on Establishment and Growth of *Atriplex halimus* and *Medicago arborea.*" *Journal of Arid Environments* 33, no. 1 (1996): 109–15.

Brahimi, Denise. *Voyageurs français du XVIIIe siècle en Barbarie.* Lille: Atelier Reproduction des Thèses, 1976.

Brice, William C., ed. *Environmental History of the Near and Middle East since the Last Ice Age.* London: Academic Press, 1978.

Brignon, Jean, Abdelaziz Amine, Brahim Boutaleb, Guy Martinet, and Bernard Rosenberger. *Histoire du Maroc.* Paris: Hatier, 1967.

Brown, John Croumbie. *French Forest Ordinance of 1669; with Historical Sketch of Previous Treatment of Forests in France.* Edinburgh: Oliver and Boyd, 1883.

Browne, Janet. *The Secular Ark: Studies in the History of Biogeography.* New Haven, Conn.: Yale University Press, 1983.

Brun, Annik. "Étude palynologique des sédiments marins holocènes de 5000 B.P. à l'actuel dans le golfe de Gabès." *Pollen et Spores* 25, no. 3–4 (1983): 437–60.

———. "Microflores et paléovégétations en Afrique du nord depuis 30,000 ans." *Bulletin de la Société Géologique de France* 8, no. 1 (1989): 25–33.

Buckley, Jerome H. *The Triumph of Time: A Study of the Victorian Concepts of Time, History, Progress, and Decadence.* Cambridge, Mass.: Belknap Press, 1966.

Burke, Edmund. "The Image of the Moroccan State in French Ethnological Literature: A New Look at the Origin of Lyautey's Berber Policy." In *Arabs and Berbers: From Tribe to Nation in North Africa,* edited by Ernest Gellner and Charles Micaud, 175–99. Lexington, Mass.: Lexington Books, 1972.

———. "La Mission scientifique au Maroc." In *Actes de Durham: Recherches récentes sur le Maroc moderne,* edited by P. Pascon, 37–56. Rabat: B.E.S.M., 1979.

———. *Prelude to Protectorate in Morocco: Precolonial Protest and Resistance, 1860–1912.* Chicago: University of Chicago Press, 1976.

Butzer, Karl W. "Environmental History in the Mediterranean World: Cross-Disciplinary Investigation of Cause-and-Effect for Degradation and Soil Erosion." *Journal of Archaeological Science* 32, no. 12 (2005): 1773–1800.

Candolle, Augustin de. "Géographie botanique." In *Ecological Phytogeography in the Nineteenth Century,* edited by Frank N. Egerton, 422–37. New York: Arno, 1977.

Canfritz, Robert, Lawrence Gowing, and David Rosand. *Places of Delight: The Pastoral Landscape.* Washington, D.C.: National Gallery of Art, 1988.

Carayol, Angel-Paul. *La Législation forestière de l'Algérie.* Paris: Arthur Rousseau, 1906.

Carette, Ernest. *Études sur la Kabilie.* Vol. 4 in the subseries Sciences historiques et géographiques of the series Exploration scientifique de l'Algérie. Paris: Imprimerie Nationale, 1848.

———. *Recherches sur l'origine et les migrations des principales tribus de l'Afrique septentrionale et particulièrement de l'Algérie.* Vol. 3. in the subseries Sciences historiques et géographiques of the series Exploration scientifique de l'Algérie. Paris: Imprimerie Impériale, 1853.

Carette, Ernest, and Auguste Warnier. *Description et division de l'Algérie.* Paris: L. Hachette, 1847.

Célérier, Jean. *Histoire du Maroc: Sommaire des leçons professées au cours de perfectionnement du service des renseignements.* Rabat: Direction des Affaires Indigènes et du Service des Renseignements, 1921.

———. *Le Maroc.* Paris: Armand Colin, 1931.

Chalvet, Martine. "Les Enjeux de la politique forestière en Algérie." *Ultramarines: Bulletin des Amis des Archives d'Outre-Mer,* no. 4 (December 1991): 13–19.

Chevalier, l'Abbé. "Deuxième note sur la flore du Sahara." *Bulletin de l'Herbier Boissier* 3, no. 1 (1903): 669–84.

Chevalier, Auguste. *L'Agronomie coloniale et le Muséum National d'Histoire Naturelle: Premières conférences du cours sur les productions coloniales végétales et l'agronomie tropicale.* Paris: Laboratoire d'Agronomie Coloniale, 1930.

Christian, P. *L'Afrique française, l'empire de Maroc et les déserts de Sahara.* Paris: A. Barbier, 1846.

Christiansen, Peter. *The Decline of Iranshahr: Irrigation and Environments in the History of the Middle East 500 B.C. to A.D. 1500.* Copenhagen: Museum Tusculanum, 1993.

Claval, Paul. "French Perceptions of Nature over the Last Centuries." Unpublished manuscript, 2005.

Clavé, Jules. "Études forestières: La Forêt de Fontainebleau." *Revue des Deux Mondes* 45 (May-June 1863): 142–70.

———. *Études sur l'économie forestière.* Paris: Guillaumin, 1862.

Cohen, Shaul E. *The Politics of Planting: Jewish-Palestinian Competition for Control of Land in the Jerusalem Periphery.* Chicago: University of Chicago Press, 1993.

———. "Promoting Eden: Tree Planting as the Environmental Panacea." *Ecumene* 6, no. 4 (1999): 424–46.

Colin, Jean. *L'Occupation romaine du Maroc. Cours préparatoire au service des affaires indigènes.* Rabat: Direction Générale des Affaires Indigènes, 1925.

Comaroff, Jean, and John Comaroff. *Of Revelation and Revolution: The Dialectics of Modernity in a South African Frontier.* Vol. 2. Chicago: University of Chicago Press, 1997.

Combe, A. D. *Les Forêts de l'Algérie.* Alger: Giralt, 1889.

———. "Notice sur les forêts de l'Algérie." In *Catalogue des collections exposées à l'Exposition Universelle et Internationale,* edited by Service des Forêts, 3–40. Alger: Adolphe Jourdan, 1885.

Condorcet, Marquis de. *Esquisse d'un tableau historique des progrès de l'esprit humain.* Paris: Agasse, 1794.

Cosson, Ernest. *Compendium florae atlanticae ou flores des états barbaresques, Algérie, Tunisie, Maroc.* Vol. 1. Paris: Imprimerie Nationale, 1881.

———. *Le Règne végétal en Algérie.* Paris: A. Quantin, 1879.

Cosson, Ernest, and Durieu de Maisonneuve. *Flore d'Algérie. Phanérogamie.* Vol. 2 in the subseries Sciences naturelles: Botanique of the series Exploration scientifique de l'Algérie. Paris: Imprimerie Impériale, 1855–67.

Coughenour, Michael B. "Graminoid Responses to Grazing by Large Herbivores: Adaptations, Exaptations, and Interacting Processes." *Annals of the Missouri Botanical Garden* 72, no. 4 (1985): 852–63.

Coughenour, Michael B., James E. Ellis, David M. Swift, D. L. Coppock, K. Galvin, J. T. McCabe, and T. C. Hart. "Energy Extraction and Use in a Nomadic Pastoral Ecosystem." *Science* 230, no. 4726 (1985): 619–25.

Cronon, William. *Changes in the Land: Indians, Colonists, and the Ecology of New England.* New Haven, Conn.: Yale University Press, 1983.

———. "A Place for Stories: Nature, History and Narrative." *Journal of American History* 78, no. 4 (March 1992): 1347–76.

Curtin, Philip D. *Death by Migration: Europe's Encounter with the Tropical World in the Nineteenth Century.* Cambridge: Cambridge University Press, 1989.

Dahl, Gudrun, and Anders Hjort. *Pastoral Herd Growth and Household Economy,* vol. 2, Stockholm Series in Social Anthropology. Stockholm: Department of Social Anthropology, 1976.

Darby, Stephen E. "Effects of Riparian Vegetation on Flow Resistance and Flood Potential." *Journal of Hydraulic Engineering* 125, no. 5 (1999): 443–54.

Davis, Diana K. "Environmentalism as Social Control? An Exploration of the Transformation of Pastoral Nomadic Societies in French Colonial North Africa." *Arab World Geographer* 3, no. 3 (2000): 182–98.

———. "Indigenous Knowledge and the Desertification Debate: Problematising Expert Knowledge in North Africa." *Geoforum* 36, no. 4 (2005): 509–24.

———. "Neoliberalism, Environmentalism and Agricultural Restructuring in Morocco." *Geographical Journal* 172, no. 2 (2006): 88–105.

———. "Potential Forests: Degradation Narratives, Science, and Environmental Policy in Protectorate Morocco, 1912–1956." *Environmental History* 10, no. 2 (2005): 211–38.

———. "Prescribing Progress: French Veterinary Medicine in the Service of Empire." *Veterinary Heritage* 29, no. 1 (2006): 1–7.

Delage, Anne-Marie. "Guillaumet, Gustave." In *Grove Art Online.* http://www.groveart.com.

Dell, Bernard, Angus J. Hopkins, and Byron B. Lamont, eds. *Resilience in Mediterranean-Type Ecosystems.* Dordrecht: Dr. W. Junk, 1986.

Delpech de Saint-Guilhem, M., M. Vialar, M. Franclieu, and M. Raousset-Boulbon. *Mémoire au roi et aux chambres, par les colons de l'Algérie.* Paris: Rignoux, 1847.

Desfontaines, René Louiche. "Fragmens d'un voyage dans les régences de Tunis et d'Alger, fait de 1785 à 1786." In *Voyages dans les régences de Tunis et d'Alger,* vol. 2, edited by M. Dureau de la Malle. Paris: Gide, 1838.

Desjobert, Amédée. *L'Algérie en 1846.* Paris: Guillaumin, 1846.

———. *L'Algérie.* Paris: Guillaumin, 1847.

Desowitz, Robert S. *The Malaria Capers: Tales of Parasites and People.* New York: W. W. Norton, 1991.

Despois, Jean. *La Tunisie.* Paris: Larousse, 1930.

Despois, Jean, and René Raynal. *Géographie de l'Afrique du nord-ouest.* Paris: Payot, 1967.

Di Castri, Francesco, David W. Goodall, and Raymond Specht, eds. *Mediterranean-Type Shrublands.* New York: Elsevier Scientific, 1981.

Dodd, Jerrold L. "Desertification and Degradation in Sub-Saharan Africa: The Role of Livestock." *BioScience* 44, no. 1 (1994): 28–34.

Doughty, Robin W. *The Eucalyptus: A Natural and Commercial History of the Gum Tree.* Baltimore: Johns Hopkins University Press, 2000.

Doumenjou, Hippolyte. *Études sur la révision du code forestier: Les Reboisements en France et en Algérie.* Paris: J. Baudry, 1883.

Dumas, Pierre. *L'Algérie.* Grenoble: B. Arthaud, 1931.

Dunn, Ross E. *Resistance in the Desert: Moroccan Responses to French Imperialism, 1881–1912.* London: Croom Helm, 1977.

Dureau de la Malle, Adolphe. *Province de Constantine: Recueil de renseigne-mens pour l'expédition ou l'établissement des français dans cette partie de l'Afrique septentrionale.* Paris: Gide, 1837.

Durrieu, Xavier. *The Present State of Morocco: A Chapter of Mussulman Civilisation.* London: Longman, Brown, Green, and Longmans, 1854.

Duval, Céleste. *Moyens de rendre aux côtes nord d'Afrique tous les principes de fécondité dont elles sont susceptibles.* Paris: Napoléon Chaix, 1859.

Duval, Jules. *L'Algérie: Tableau historique, descriptif et statistique.* Paris: L. Hachette, 1859.

Duval, Jules, and Auguste Warnier. *Un Programme de politique algérienne.* Paris: Challamel Ainé, 1868.

Egler, Frank E. "Philosophical and Practical Considerations of the Braun-Blanquet System of Phytosociology." *Castanea* 19, no. 2 (1954): 45–60.

Elenga, H., O. Peyron, R. Bonnefille, D. Jolly, R. Cheddadi, J. Guiot, V. Andrieu et al. "Pollen-Based Biome Reconstruction for Southern Europe and Africa 18,000 yr BP." *Journal of Biogeography* 27, no. 3 (2000): 11–34.

El Ghonemy, Mohammad Riad. *Land, Food, and Rural Development in North Africa.* Westview Special Studies in Social, Political, and Economic Development. Boulder: Westview, 1993.

Ellis, James E., and David M. Swift. "Stability of African Pastoral Ecosystems: Alternate Paradigms and Implications for Development." *Journal of Range Management* 41, no. 6 (1988): 450–59.

Emberger, Louis. "Aperçu général [de la végétation]." In *La Science au Maroc,* edited by Paul Boudy and P. Despujols, 149–82. Casablanca: Imprimeries Réunies, 1934.

———. "Aperçu général sur la végétation du Maroc. Commentaire de la carte phytogéographique du Maroc." *Veröffentlichungen des Geobotanischen Institutes Rübel in Zürich* 14, no. 1 (1939): 40–157.

———. "Charles Flahault (1852–1935)." *Revue Générale de Botanique* 48, no. 565 (1936): 1–48.

———. "L'Importance du chêne-liège dans le paysage marocain avant la destruction des forêts." *La Revue de Géographie Marocaine* 7, no. 1 (1928): 121–24.

———. Préface to *Flore de l'Afrique du nord,* edited by René Maire, 1–6. Paris: Paul Lechevalier, 1952.

———. *Travaux de botanique et d'écologie.* Paris: Masson, 1971.

———. "La Végétation de la région méditerranéenne: Essai d'une classification des groupements végétaux." *Revue Générale de Botanique* 42, no. 503 (1930): 641–62 and 705–21.

Endfield, Georgina H., and David J. Nash. "Drought, Desiccation and Discourse: Missionary Correspondence and Nineteenth-Century Climate Change in Central Southern Africa." *Geographical Journal* 168, no. 1 (2002): 33–47.

Enfantin, Prosper. *Colonisation de l'Algérie.* Paris: P. Bertrand, 1843.

Ettori, Commandant. "Le Berbère et le code forestier." CHEAM: unpublished report, 1952.

Evelyn, John. *Silva: Or, A Discourse of Forest-Trees.* York: A. Ward, 1776.

Fabre, Jean Antoine. *Essai sur la théorie des torrens et des rivières.* Paris: Bidault, 1797.

Fairhead, James, and Melissa Leach. *Misreading the African Landscape: Society and Ecology in a Forest-Savanna Mosaic.* Cambridge: Cambridge University Press, 1996.

———. *Reframing Deforestation: Global Analyses and Local Realities; Studies in West Africa.* London: Routledge, 1998.

Faucon, Narcisse. *Le Livre d'or de l'Algérie: Histoire politique, militaire, administrative, événements et faits principaux, biographie des hommes ayant marqué dans l'armée, les sciences, les lettres, etc.* Paris: Challamel, 1889.

Faust, D., C. Zielhofer, R. B. Escudero, and F. D. del Olmo. "High-Resolution Fluvial Record of Late Holocene Geomorphic Change in Northern Tunisia: Climate or Human Impact?" *Quaternary Science Reviews* 23, no. 16–17 (2004): 1757–75.

Fennane, Mohamed, ed. *La Grande encyclopédie du Maroc: Flore et végétation.* Rabat: GEI, 1987.

Ficheur, E. "Relation entre la constitution géologique du sol et la distribution des boisements." In *Les Forêts de l'Algérie,* edited by Henri Lefebvre, 70–91. Alger-Mustapha: Giralt, 1900.

Figueroa, M. E., and A. J. Davy. "Response of Mediterranean Grassland Species to Changing Rainfall." *Journal of Ecology* 79, no. 4 (1991): 925–41.

Filias, Achilles. *Notice sur les forêts de l'Algérie: Leur étendue; leurs essences; leurs produits.* Alger: Gojosso, 1878.

Flahault, Charles. "Au Sujet de la carte botanique, forestière et agricole de France." *Annales de Géographie* 5, no. 15 (October 1896): 449–57.

Food and Agriculture Organization of the United Nations. *Forest Resources Assessment 1990: Non-Tropical Developing Countries Mediterranean Region.* Rome: FAO, 1994.

Ford, Caroline. "Landscape and Environment in French Historical and Geographical Thought: New Directions." *French Historical Studies* 24, no. 1 (2001): 125–34.

————. "Nature, Culture and Conservation in France and Her Colonies 1840–1940." *Past and Present* 183, no. 1 (2004): 173–98.

Franc, Julien. *La Colonisation de la Mitidja.* Paris: Honoré Champion, 1928.

Freeman, John F. "Forest Conservancy in the Alps of Dauphiné, 1287–1870." *Forest and Conservation History* 38 (October 1994): 171–80.

Frémeaux, Jacques. "Souvenirs de Rome et présence française au Maghreb: Essai d'investigation." In *Connaissances du Maghreb: Sciences sociales et colonisation,* edited by Jean-Claude Vatin, 29–46. Paris: CNRS, 1984.

Fromentin, Eugène. *Between Sea and Sahara: An Algerian Journal.* Translated by Blake Robinson. Athens: Ohio University Press, 1999.

————. *Sahara et Sahel. I.—Une été dans le Sahara. II.—Une année dans le Sahel.* Illustrated ed. Paris: E. Plon, 1879.

Gadiou, Vétérinaire-Major. "La Situation économique du Sous." *Revue de Géographie du Maroc* 6, no. 1 (1927): 137–65.

Galibert, Léon. *L'Algérie ancienne et moderne.* Paris: Furne et Cie., 1844.

Gautier, Émile F. *Le Passé de l'Afrique du nord: Les Siècles obscurs.* Paris: Payot, 1942.

Germain, José, and Stéphane Faye. *Le Nouveau monde français: Maroc—Algérie—Tunisie.* Paris: Plon, 1924.

Glacken, Clarence. *Traces on the Rhodian Shore.* Berkeley: University of California Press, 1967.

Glucq. *L'Album de l'exposition.* Paris: Ch. Gaulon, 1889.

Godlewska, Anne. "Map, Text and Image. The Mentality of Enlightened Conquerors: A New Look at the *Description de l'Egypte.*" *Transactions of the Institute of British Geographers* 20, no. 1 (1995): 5–28.

Goldschmidt, Walter. "The Failure of Pastoral Economic Development Projects in Africa." In *The Future of Pastoral Peoples,* edited by John Galaty, 101–18. Ottowa: International Development Research Center, 1981.

Goudie, Andrew. *The Human Impact on the Natural Environment.* Cambridge, Mass.: MIT Press, 2000.

Gouvernement Général de l'Algérie. *Commission de législation forestière de l'Algérie: Projet de loi.* Alger: Giralt, 1893.

————. *Commission d'études forestières.* Alger: J. Torrent, 1904.

————. *Création d'un service pastoral.* Alger-Mustapha: Giralt, 1896.

————. *Programme général du reboisement.* Alger: Imprimerie Administrative, 1885.

Grove, A. T., and Oliver Rackham. *The Nature of Mediterranean Europe: An Ecological History.* New Haven, Conn.: Yale University Press, 2001.

Grove, Richard H. *Ecology, Climate, and Empire: Colonialism and Global Environmental History, 1400–1940.* Cambridge: White Horse, 1997.

———. *Green Imperialism: Colonial Expansion, Tropical Island Edens, and the Origins of Environmentalism, 1600–1860.* Cambridge: Cambridge University Press, 1995.

———. "Scottish Missionaries, Evangelical Discourses and the Origins of Conservation Thinking in Southern Africa, 1820–1900." *Journal of Southern African Studies* 15, no. 2 (1989): 163–87.

Grove, Richard H., Vinita Damodaran, and Satpal Sangwan, eds. *Nature and the Orient: The Environmental History of South and Southeast Asia.* Delhi: Oxford University Press, 1998.

Gsell, Stéphane. *Les conditions du développement historique, les temps primitifs, la colonisation phénicienne et l'empire de Carthage.* Vol. 1 of *Histoire ancienne de l'Afrique du nord.* 3rd ed. Paris: Hachette, 1921.

Guha, Ramachandra. *The Unquiet Woods: Ecological Change and Peasant Resistance in the Himalaya.* New Delhi: Oxford University Press, 1989.

Guinier, Philibert. "Nécrologie: Paul Boudy (1874–1957)." *Revue Forestière Française* 3 (May 1958): 219–22.

———. "René Maire, 1878–1949, In Memoriam." *Bulletin de la Société d'Histoire Naturelle de l'Afrique du Nord* 41, no. 1 (1950): 64–113.

Gutterman, Yitzchak. *Regeneration of Plants in Arid Ecosystems Resulting from Patch Disturbance.* Dordrecht: Kluwer Academic, 2001.

Guyot, Charles. *L'Enseignement forestier en France: L'École de Nancy.* Nancy: Crépin-Leblond, 1898.

Haase, Peter, Francisco Pugnaire, and L. D. Incoll. "Seed Production and Dispersal in the Semi-arid Tussock Grass *Stipa tenacissima* L. during Masting." *Journal of Arid Environments* 31, no. 1 (1995): 55–65.

Hannoum, Abdelmajid. "Translation and the Colonial Imaginary: Ibn Khaldûn Orientalist." *History and Theory* 42, no. 1 (February 2003): 61–81.

Hardy, Auguste. "Note climatologique sur l'Algérie au point de vue agricole." In *Recueil de traités d'agriculture et d'hygiène à l'usage des colons de l'Algérie,* edited by Ministre de la Guerre, 37–62. Alger: Imprimerie du Gouvernement, 1851.

Harrison, Robert P. *Forests: The Shadow of Civilization.* Chicago: University of Chicago Press, 1992.

Hastings, Alan, Carole Hom, Stephen Ellner, Peter Turchin, and H. Charles Godfray. "Chaos in Ecology: Is Mother Nature a Strange Attractor?" *Annual Review of Ecology and Systematics* 24 (1993): 1–33.

Heffernan, Michael J. "The Desert in French Orientalist Painting during the

Nineteenth Century." *Landscape Research* 16, no. 2 (1991): 37–42.

———. "An Imperial Utopia: French Surveys of North Africa in the Early Colonial Period." In *Maps and Africa*, edited by J. Stone, 81–107. Aberdeen: Aberdeen University African Studies Group, 1994.

Heffernan, Michael, and Keith Sutton. "The Landscape of Colonialism: The Impact of French Colonial Rule on the Algerian Rural Settlement Pattern, 1830–1987." In *Colonialism and Development in the Contemporary World*, edited by Christopher Dixon and Michael Heffernan, 121–52. London: Mansell, 1991.

Heusch, Bernard. "Cinquante ans de banquettes de DRS-CES en Afrique du nord: Un bilan." *Cahiers ORSTROM* 22, no. 2 (1986): 153–62.

Housefield, James E. "Orientalism as Irony in Gérard de Nerval's *Voyage en Orient*." *Journal of North African Studies* 5, no. 4 (2000): 10–24.

Huffel, Gustave. *Économie forestière, tome premier, deuxième partie: Propriété et législation forestières, politique forestière, la France forestière, statistiques*. 2nd ed. Paris: Librairie Agricole de la Maison Rustique, 1920.

Ibn Khaldoun [Khaldûn]. *Histoire de l'Afrique sous la dynastie des aghlabites, et de la Sicile sous la domination musulmane*. Translated by Adolphe-Noël Desvergers. Paris: F. Didot Frères, 1841.

———. *Histoire des berbères et des dynasties musulmanes de l'Afrique septentrionale*. 4 vols. Translated by William MacGuckin de Slane. Alger: Imprimerie du Gouvernement, 1852–56.

———. *Histoire des berbères et des dynasties musulmanes de l'Afrique septentrionale*. Vol. 1. Translated by William MacGuckin de Slane. Edited by Paul Casanova. Paris: Librairie Orientaliste Paul Geuthner, 1925.

———. *The Muqaddimah: An Introduction to History*. Translated by Franz Rosenthal. Edited by N. J. Dawood. Princeton, N.J.: Princeton University Press, 1967.

Idris, Roger. "De la réalité de la catastrophe hilâlienne." *Annales: Économies, Sociétés, Civilisations* 23, no. 2 (1968): 390–96.

Idrisi. *Description de l'Afrique et de l'Espagne*. Translated by Reinhart P. A. Dozy and Michel Jean de Goeje. Leiden: E. J. Brill, 1968 [1866].

Ilahiane, Hsain. "The Berber Agdal Institution: Indigenous Range Management in the Atlas Mountains." *Ethnology* 38, no. 1 (1999): 21–46.

Institut du Monde Arabe. *De Delacroix à Renoir: L'Algérie des peintres*. Paris: Institut du Monde Arabe/Hazan, 2003.

Jacoby, Karl. *Crimes against Nature: Squatters, Poachers, Thieves, and the Hidden History of American Conservation*. Berkeley: University of California Press, 2001.

Jacquot, André. *Incendies en forêt (Forest Fires)*. Translated by C. E. C. Fischer. Calcutta: Superintendant Government Printing, 1910.

Jourde, J. *La Ligue pour la défense de la colonisation sur les hauts-plateaux à monsieur Laferrière, gouverneur général de l'Algérie*. Saida: Émile Pascal, 1898.

Julien, Charles-André. *Histoire de l'Algérie contemporaine: La Conquête et les débuts de la colonisation (1827–1871)*. Paris: Presses Universitaires de France, 1964.

Kaufmann, Jeffrey. "The Cactus Was Our Kin: Pastoralism in the Spiny Desert of Southern Madagascar." In *Changing Nomads in a Changing World*, edited by Joseph Ginat and Anatoly Khazanov, 124–42. Brighton: Sussex Academic, 1998.

Kipling, Rudyard. "In the Rukh." In *Many Inventions*, 222–64. New York: D. Appleton, 1899.

Knight, Melvin. *Morocco as a French Economic Venture: A Study of Open Door Imperialism*. New York: D. Appleton-Century, 1937.

Kull, Christian. "Deforestation, Erosion, and Fire: Degradation Myths in the Environmental History of Madagascar." *Environment and History* 6, no. 4 (2000): 423–50.

Lacoste, Yves. *Ibn Khaldun: The Birth of History and the Past of the Third World*. Translated by David Macey. London: Verso, 1984.

Lacoste, Yves, André Nouschi, and André Prenant. *L'Algérie: Passé et présent*. Paris: Éditions Sociales, 1960.

Lahlimi Alami, Ahmed, ed. *La Grande encyclopédie du Maroc: Agriculture et pêche*. Rabat: GEI, 1987.

Lamb, H. F., F. Damblon, and R. W. Maxted. "Human Impact on the Vegetation of the Middle Atlas, Morocco, during the Last 5000 Years." *Journal of Biogeography* 18, no. 5 (1991): 519–32.

Lamb, H. F., C. A. Duigan, J. H. Gee, and K. Kelts. "Lacustrine Sedimentation in a High-Altitude, Semi-arid Environment: The Palaeolimnological Record of Lake Isli, High Atlas, Morocco." In *Environmental Change in Drylands: Biogeographical and Geomorphological Perspectives*, edited by Andrew C. Millington and Ken Pye, 147–61. New York: John Wiley, 1994.

Lamb, H. F., U. Eichner, and V. R. Switsur. "An 18,000-Year Record of Vegetation, Lake-Level and Climate Change from Tigalmamine, Middle Atlas, Morocco." *Journal of Biogeography* 1, no. 16 (1989): 65–74.

Lapasset, Ferdinand. *Aperçu sur l'organisation des indigènes dans les territoires militaires et dans les territoires civils*. Alger: Dubos Frères, 1850.

Lapène, Édouard. *Vingt-six mois à Bougie ou collection de mémoires sur sa conquête, son occupation et son avenir*. Paris: Bouchene, 2002.

Lavauden, Louis. "Les Forêts du Sahara." *Revue des Eaux et Forêts* 65, no. 6 (1927): 265–77 and 329–41.

Leach, Melissa, and Robin Mearns, eds. *The Lie of the Land: Challenging Received Wisdom on the African Environment*. London: International African Institute, 1996.

Lefebvre, Henri. *Les Forêts de l'Algérie*. Alger-Mustapha: Giralt, 1900.

———. *Notice sur les forêts de la Tunisie et catalogue raisonné*. Tunis: B. Borrel, 1889.

Lefranc, Edmond. "Sidi-Bel-Abbès: Topographie, Climatologie et Botanique." *Bulletin de la Société Botanique de France* 12 (1865): 383–95 and 415–31.

Le Houérou, Henry N. "Man-Made Deserts: Desertization Processes and Threats." *Arid Land Research and Management* 16, no. 1 (2002): 1–36.

———. "Restoration and Rehabilitation of Arid and Semiarid Mediterranean Ecosystems in North Africa and West Asia: A Review." *Arid Soil Research and Rehabilitation* 14, no. 1 (2000): 3–14.

Lehureaux, Léon. "Comment s'effectue actuellement la transhumance." *L'Union Ovine* 2, no. 5 (1930): 189–91.

———. *Le Nomadisme et la colonisation dans les hauts plateaux de l'Algérie*. Paris: Comité de l'Afrique Française, 1931.

Ligue du Reboisement. *La Forêt: Conseils aux indigènes*. Alger: Association Ouvrière P. Fontana, 1883.

Lopez-Vera, Fernando, ed. *Quaternary Climate in Western Mediterranean*. Madrid: Universidad Autónoma de Madrid, 1986.

Lorcin, Patricia M. *Imperial Identities: Stereotyping, Prejudice and Race in Colonial Algeria*. London: I. B. Tauris, 1995.

———. "Rome and France in Africa: Recovering Colonial Algeria's Latin Past." *French Historical Studies* 25, no. 2 (2002): 295–329.

Lorin, Henri. *L'Afrique du nord: Tunisie, Algérie, Maroc*. Paris: Armand Colin, 1908.

Lyautey, Louis. *Paroles d'action*. Paris: Imprimerie Nationale, 1995.

———, ed. *Rapport général sur la situation du protectorat du Maroc*. Rabat: Résidence Générale de la République Française au Maroc, 1916.

Mac Carthy, M. Oscar. *Géographie physique, économique et politique de l'Algérie*. Alger: Dubos Frères, 1858.

Mack, Richard, and John Thompson. "Evolution in Steppe with a Few Large, Hooved Mammals." *American Naturalist* 119, no. 6 (1982): 757–773.

Maire, René. *Carte phytogéographique de l'Algérie et de la Tunisie: Notice*. Alger: Baconnier Frères, 1926.

————. "Coup d'oeil sur la végétation du Maroc." In *Sur les productions végétales du Maroc,* edited by Émile Perrot and Louis Gentil, 59–71. Paris: Larose, 1921.

————. *Flore de l'Afrique du nord (Maroc, Algérie, Tunisie, Tripolitaine, Cyrénaique et Sahara).* Paris: Paul Lechevalier, 1952.

————. "Louis Trabut." *Revue de Botanique Appliquée et d'Agriculture Tropicale* 9, no. 98 (1929): 613–20.

————. *Les Progrés des connaissances botaniques en Algérie depuis 1830.* Paris: Masson, 1931.

Maire, René, and Charles-Louis Trabut. "Jules-Aimé Battandier (1848–1922)." *Bulletin de la Société Botanique de France* 23, no. 23 (February 1923): 106–17.

Mangenot, G. "La Carrière et l'oeuvre." In *Travaux de botanique et d'écologie,* by Louis Emberger, 1–9. Paris: Masson, 1971.

Marc, Henri. *Notes sur les forêts de l'Algérie.* Alger: Adolphe Jourdan, 1916.

————. *Notes sur les forêts de l'Algérie.* Collection du centenaire de l'Algérie, 1830–1930. Paris: Larose, 1930.

Marsh, George Perkins. *Man and Nature: Or, Physical Geography as Modified by Human Action.* Edited by David Lowenthal. Cambridge, Mass.: Belknap Press, 1965.

Martin, L.-A. "L'Élevage au Maroc." *Maroc Médical* 28, no. 292 (1949): 468–79.

Matagne, Patrick. *Aux Origines de l'écologie: Les Naturalistes en France de 1800 à 1914.* Paris: CTHS, 1999.

Mathieu, Auguste. *Flore forestière: Description et histoire des végétaux ligneux qui croissent spontanément en France et des essences importantes de l'Algérie.* 2nd ed. Nancy: Ancienne Maison Grimblot, 1860.

Mathieu, Auguste, and Louis Trabut. *Les Hauts-plateaux Oranais, rapport de mission.* Alger: Pierre Fontana, 1891.

Maury, Louis Ferdinand Alfred. *Histoire des grandes forêts de la Gaule et de l'ancienne France.* Paris: A. Leleux, 1850.

McCann, James. *Green Land, Brown Land, Black Land: An Environmental History of Africa.* Portsmouth, N.H.: Heinemann, 1999.

McIntosh, Robert. "Pluralism in Ecology." *Annual Review of Ecology and Systematics* 18 (1987): 321–41.

McNeill, John R. *The Mountains of the Mediterranean World: An Environmental History.* New York: Cambridge University Press, 1992.

————. "Observations on the Nature and Culture of Environmental History." *History and Theory* 42, no. 4 (December 2003): 5–43.

Messerli, Bruno, and Matthias Winiger. "Climate, Environmental Change, and Resources of the African Mountains from the Mediterranean to the Equator." *Mountain Research and Development* 12, no. 4 (1992): 315–36.

Metro, André. *Forêts, Atlas du Maroc, notices explicatives, section VI, biogéographie: Forêts et ressources végétales.* Rabat: Comité de Géographie du Maroc, 1958.

Miège, Jacques. "In Memoriam: Louis Emberger." *Candollea* 25, no. 2 (1970): 183–87.

Mikesell, Marvin. "Deforestation in Northern Morocco." *Science* 132, no. 3425 (1960): 441–48.

Milchunas, Daniel G., and William K. Lauenroth. "Quantitative Effects of Grazing on Vegetation and Soils over a Global Range of Environments." *Ecological Applications* 63, no. 4 (1993): 327–66.

Mitchell, B. R. *International Historical Statistics, Africa and Asia.* New York: New York University Press, 1982.

Moll, Louis. *Colonisation et agriculture de l'Algérie.* Paris: Librairie Agricole de la Maison Rustique, 1845.

Monod, Théophile. "L'Élevage au Maroc." In *La Renaissance du Maroc: Dix ans de protectorat,* edited by Résidence Générale de la République Française au Maroc, 294–306. Rabat: Résidence Générale de la République Française au Maroc, 1924.

———. *L'Élevage au Maroc: Cours préparatoire au service des affaires indigènes.* Rabat: Résidence Générale de France au Maroc, Direction Générale des Affaires Indigènes, 1927.

———. "Le Rôle du vétérinaire colonial." *La Terre Marocaine* 1, no. 4 (1931): 4–15.

Munby, G. *Flore de l'Algérie ou catalogue des plantes indigènes du royaume d'Alger.* Paris: J.-B. Baillière, 1847.

Murbeck, Svante S. *Contributions à la connaissance de la flore du nord-ouest de l'Afrique et plus spécialement de la Tunisie.* Lund: E. Malmstrom, 1897.

———. *Contributions à la connaissance de la flore du nord-ouest de l'Afrique et plus spécialement de la Tunisie, deuxième série.* Lund: Hakan Ohlsson, 1905.

Nahal, I. "The Mediterranean Climate from a Biological Viewpoint." In *Mediterranean-Type Shrublands,* edited by Francesco di Castri, David W. Goodall, and Raymond Specht, 63–86. New York: Elsevier Scientific, 1981.

Naveh, Zev. "The Role of Fire and Its Management in the Conservation of Mediterranean Ecosystems and Landscapes." In *The Role of Fire in Mediterranean-Type Ecosystems,* edited by José Moreno and Walter Oechel, 163–85. New York: Springer-Verlag, 1994.

Nelson, Harold. *Algeria: A Country Study.* Washington, D.C.: American University, 1985.

———. *Morocco: A Country Study.* Washington, D.C.: American University, 1985.

———. *Tunisia: A Country Study.* Washington, D.C.: American University, 1988.

Neumann, Roderick P. *Imposing Wilderness: Struggles over Livelihood and Nature Preservation in Africa.* Berkeley: University of California Press, 1998.

Niamir-Fuller, Maryam. "The Resilience of Pastoral Herding in Sahelian Africa." In *Linking Social and Ecological Systems: Management Practices and Social Mechanisms,* edited by Fikret Berkes and Carl Folke, 250–84. Cambridge: Cambridge University Press, 1998.

Nicholson, Sharon E. "Climatic Variations in the Sahel and Other African Regions during the Last Five Centuries." *Journal of Arid Environments* 1, no. 11 (1978): 3–24.

———. "The Historical Climatology of Africa." In *Climate and History,* edited by T. Wigley, M. Ingram, and G. Farmer, 249–70. Cambridge: Cambridge University Press, 1981.

———. "Recent Rainfall Fluctuations in Africa and Their Relationship to Conditions over the Continent." *Holocene* 4, no. 2 (1994): 121–31.

Nouschi, André. *Enquête sur le niveau de vie des populations rurales constantinoises de la conquête jusqu'en 1919.* Paris: Presses Universitaires de France, 1961.

———. "Notes sur la vie traditionelle des populations forestières algériennes." *Annales de Géographie* 68, no. 370 (1959): 525–35.

Noyes, John. "Nomadic Fantasies: Producing Landscapes of Mobility in German Southwest Africa." *Ecumene* 7, no. 1 (2000): 47–66.

Noy-Meir, Imanuel, and Eddy van der Maarel. "Relations between Community Theory and Community Analysis in Vegetation Science: Some Historical Perspectives." *Vegetatio* 69, no. 1–3 (1987): 5–15.

Oba, Gufu, Nils Stenseth, and Walter Lusigi. "New Perspectives on Sustainable Grazing Management in Arid Zones of Sub-Saharan Africa." *BioScience* 50, no. 1 (2000): 35–51.

Olsvig-Whittaker, Linda, Eliezer Frankenberg, Avi Perevolotsky, and Eugene D. Ungar. "Grazing, Overgrazing, and Conservation: Changing Concepts and Practices in the Negev Rangelands. *Sécheresse* 17, no. 1–2 (2006): 195–99.

O'Rourke, Eileen. "Changing Identities, Changing Landscapes: Human-Land Relations in Transition in the Aspre, Roussillon." *Ecumene* 6, no. 1 (1999): 29–50.

Osborne, Michael A. *Nature, the Exotic, and the Science of French Colonialism.* Bloomington: Indiana University Press, 1994.

Ouaskioud, Driss. "Contribution à l'étude de la dynamique de la végétation steppique après une mise en défens de longue durée: Cas de la station d'amélioration pastorale Anbad Boumalne Dades (Ouarzazate)." Mémoire de Troisième Cycle, Institut Agronomique et Vétérinaire Hassan II, 1999.

Parker, Harold. *The Cult of Antiquity and the French Revolutionaries.* Chicago: University of Chicago Press, 1937.

Pellissier de Reynaud, E. *Mémoires historiques at géographiques sur l'Algérie.* Vol. 6 in the subseries Sciences historiques et géographiques of the series Exploration scientifique de l'Algérie. Paris: Imprimerie Royale, 1844.

Peltre, Christine. *Orientalism in Art.* Translated by John Goodman. New York: Abbeville, 1998.

Peluso, Nancy L. *Rich Forests, Poor People: Resource Control and Resistance in Java.* Berkeley: University of California Press, 1992.

Pepper, Edward. "Eucalyptus in Algeria and Tunisia, from an Hygienic and Climatological Point of View." *Proceedings of the American Philosophical Society* 35, no. 150 (1896): 39–56.

Perevolotsky, Avi, and No'am Seligman. "Role of Grazing in Mediterranean Rangeland Ecosystems." *BioScience* 48, no. 12 (1998): 1007–17.

Périer, J.-A.-N. *De l'hygiène en Algérie.* Vol. 1 in the subseries Sciences médicales of the series Exploration scientifique de l'Algérie. Paris: Imprimerie Royale, 1847.

Perkins, Kenneth J. *A History of Modern Tunisia.* New York: Cambridge University Press, 2004.

Peyssonnel, Jean-André. "Relation d'un voyage sur les côtes de la Barbarie fait par ordre du roi en 1724 et 1725." In *Voyages dans les régences de Tunis et d'Alger,* vol. 1, edited by M. Dureau de la Malle. Paris: Gide, 1838.

Pfeifer, Karen. "The Development of Commercial Agriculture in Algeria, 1830–1970." *Research in Economic History* 10 (1986): 277–308.

Pitard, M. C.-J. *Exploration scientifique du Maroc: Premier fascicule: Botanique.* Paris: Masson, 1913.

Plaisance, Georges. *La Forêt française.* Paris: Édition Denoël, 1979.

Planchon, J. E. "L'Eucalyptus globulus au point de vue botanique, économique et médical." *Revue des Deux Mondes* 7, no. 1 (January 1875): 149–74.

Planhol, Xavier de. *Les Fondements géographiques de l'histoire de l'Islam.* Paris: Flammarion, 1968.

Plateau, Henri. "La Lutte contre le ruissellement." *Bulletin de la Société des Sciences Naturelles du Maroc* 28, no. 1 (1948): 100–9.

Pliny. *Natural History.* Vol. 2. *Books 3–7.* Translated by H. Rackham. Cambridge, Mass.: Harvard University Press, 1942.

Plit, Florian. "La Dégradation de la végétation, l'érosion et la lutte pour protéger le milieu naturel en Algérie et au Maroc." *Méditerranée* 49, no. 3 (1983): 79–88.

Poncet, Jean. "Le Mythe de la 'catastrophe' Hilalienne." *Annales: Économies, Sociétés, Civilisations* 22, no. 5 (1967): 1099–1120.

———. *Paysages et problèmes ruraux en Tunisie.* Paris: Presses Universitaires de France, 1962.

Poncet, Jean, and André Raymond. *La Tunisie.* Paris: Presses Universitaires de France, 1971.

Pons, Armand, and Pierre Quezel. "The History of the Flora and Vegetation and Past and Present Human Disturbance in the Mediterranean Region." In *Plant Conservation in the Mediterranean Area,* edited by César Gomez-Campo, 25–43. Dordrecht: Dr. W. Junk, 1985.

Price, Roger. *A Concise History of France.* Cambridge: Cambridge University Press, 1993.

Puyo, Jean-Yves. "Lyautey et la politique forestière marocaine (Protectorat français, 1912–1956)." Paper presented at Géographie, Exploration et Colonisation (XIXe-XXe siècles), Colloque Bordeaux, 13–15 October 2005.

Pyne, Stephen J. *Vestal Fire: An Environmental History, Told through Fire, of Europe and Europe's Encounter with the World.* Seattle: University of Washington Press, 1997.

Rabot, Charles. "Le Déboisement et l'aggravation de la torrentialité en Algérie." *La Géographie, Bulletin de la Société de Géographie* 11, no. 15 (July 1905): 315–17.

Raguse, duc de. *Voyage du maréchal duc de Raguse.* Paris: Ladvocat, 1837.

Rajan, S. Ravi. *Modernizing Nature: Forestry and Imperial Eco-Development, 1800–1950.* Oxford: Clarendon, 2006.

Raousset-Boulbon, comte de. *De la colonisation et des institutions civiles en Algérie.* Paris: Dauvin et Fontaine, 1847.

Raup, Hugh M. "Trends in the Development of Geographic Botany." *Annals of the Association of American Geographers* 32, no. 4 (1942): 319–54.

Raynal, G. T. *Histoire philosophique et politique des établissements et du commerce des Européens dans l'Afrique septentrionale.* 2 vols. Paris: Amable Costes, 1826.

Reale, Oreste, and Paul Dirmeyer. "Modeling Effects of Vegetation on Mediterranean Climate during the Roman Classical Period." *Global and Planetary Change* 25, no. 3–4 (2000): 163–84.

Reille, Maurice. "Contribution pollenanalytique à l'histoire holocène de la végétation des montagnes du Rif (Maroc septentrional)." *Supplément au Bulletin AFEQ* 1, no. 50 (1977): 53–76.

Renaudot, M. *Alger: Tableau du royaume, de la ville d'Alger et de ses environs.* 4th ed. Paris: P. Mongie Aîné, 1830.

République Algérienne Democratique et Populaire. Direction des Forêts. *Rapport national de l'Algérie sur la mise en oeuvre de la convention de lutte contre la désertification.* Alger: Ministère de l'Agriculture et du Developpement Rural, Direction des Forêts, 2004.

Résidence Générale de France à Tunis. Service des Affaires Indigènes. *Historique de l'annexe des affaires indigènes de Ben-Gardane.* Bourg: Victor Berthod, 1931.

———. Service des Affaires Indigènes. *Historique du bureau des affaires indigènes de Matmata.* Bourg: Victor Berthod, 1931.

———. Service des Forêts. *Recueil des décrets et arrétés constituant la législation forestière tunisienne.* Tunis: Imprimerie Centrale, 1936.

Reynard, J. *Restauration des forêts et des pâturages du sud de l'Algérie (province d'Alger).* Alger: Adolphe Jourdan, 1880.

Ritchie, James C. "Analyse pollinique de sédiments holocènes supérieurs des hauts plateaux du Maghreb oriental." *Pollen et Spores* 26, no. 3–4 (1984): 489–96.

Rivet, Daniel. *Lyautey et l'instauration du protectorat français au Maroc, 1912–1925.* 3 vols. Paris: Harmattan, 1988.

Rivière, Charles, and H. Lecq. *Traité pratique d'agriculture pour le nord de l'Afrique.* Paris: Société d'Éditions, 1928.

Robbins, Paul. "Fixed Categories in a Portable Landscape: The Causes and Consequences of Land-Cover Categorization." *Environment and Planning A* 33, no. 1 (2001): 161–79.

Roberts, Neil. *The Holocene: An Environmental History.* Oxford: Blackwell, 1998.

Rodes, John E. *A Short History of the Western World.* New York: Charles Scribner's Sons, 1970.

Rognon, P. "Late Quaternary Climatic Reconstruction for the Maghreb (North Africa)." *Palaeogeography, Palaeoclimatology, Palaeoecology* 58, no. 1–2 (1987): 11–34.

Rolet, Antonin. "La Transhumance en France: Ses difficultés." *L'Union Ovine* 2, no. 9 (1930): 386–89.

Rondet-Saint, Maurice. "Un Problème à résoudre en Afrique du nord: Le Nomadisme." *La Dépêche Coloniale,* 18 December 1931, 1.

Ruedy, John. *Modern Algeria: The Origins and Development of a Nation.* Bloomington: Indiana University Press, 1992.

Ruet, Commandant. "La Transhumance dans le moyen Atlas et la haute Moulouya." CHEAM: unpublished report, 1952.

Saberwal, Vasant. "Science and the Desiccationist Discourse of the 20th Century." *Environment and History* 4, no. 3 (1998): 309–43.

Sahlins, Peter. *Forest Rites: The War of the Demoiselles in Nineteenth-Century France.* Cambridge, Mass.: Harvard University Press, 1994.

Saint-Germes, Jean. *Économie algérienne.* Alger: La Maison des Livres, 1955.

———. *La Réforme agraire algérienne.* Alger: La Maison des Livres, 1955.

Salamani, M. "Premières données paléophytogéographiques du cèdre de l'Atlas (*Cedrus atlanticus*) dans la région de grande Kabylie (N-E Algérie)." *Palynosciences* 2 (1993): 147–55.

Salvy, Capitaine. "La Crise du nomadisme dans le sud marocain." CHEAM: unpublished report, 1949.

Sari, Djilali. *La Dépossession des fellahs.* Alger: Société Nationale d'Édition et de Diffusion, 1978.

Sauvage, Charles. "La Forêt marocaine." *Bulletin de l'Enseignement Publique* 28, no. 169 (1941): 235–57.

Schoenenberger, A. *Utilisation de la végétation naturelle dans la vallée du Dra: Lutte contre la désertification par une amélioration sylvo-pastorale.* Zagora: PROLUDRA/ORMVA, 1994.

Scoones, Ian. "Range Management Science and Policy: Politics, Polemics and Pasture in Southern Africa." In *The Lie of the Land: Challenging Received Wisdom on the African Environment,* edited by Melissa Leach and Robin Mearns, 34–53. London: International African Institute, 1996.

Scoones, Ian, ed. *Living with Uncertainty: New Directions in Pastoral Development in Africa.* London: Intermediate Technology Publications, 1995.

Service de l'Élevage du Maroc. "Historique du service de l'élevage du Maroc: Ses attributions et son rôle dans l'économie du pays." *Maroc Médical* 28, no. 292 (1949): 471–74.

Shaler, William. *Esquisse de l'état d'Alger considéré sous les rapports politique, historique et civil.* Translated by Thomas-Xavier Bianchi. Paris: Ladvocat, 1830.

———. *Sketches of Algiers, Political, Historical and Civil, Containing an Account of the Geography, Population, Government, Revenues, Commerce, Agriculture, Arts, Civil Institutions, Tribes, Manners, Language and Recent Political History of That Country.* Boston: Cummings, Hilliard, 1826.

Shaw, Brent. "Climate, Environment, and History: The Case of Roman North Africa." In *Climate and History: Studies in Past Climates and their Impact on Man,* edited by T. Wigley, M. Ingram, and G. Farmer, 379–403. Cambridge: Cambridge University Press, 1981.

Shaw, Thomas. "Travels or Observations Relating to Barbary." In *A General Collection of the Best and Most Interesting Voyages and Travels in All Parts of the World,* edited by John Pinkerton, 499–680. London: Longman, Hurst, Rees, Orme, and Brown, 1814.

———. *Travels or Observations Relating to Several Parts of Barbary and the Levant.* Oxford: Printed at the Theatre, 1738.

Showers, Kate B. "Gallery: Kate B. Showers on Mapping African Soils." *Environmental History* 10, no. 2 (2005): 314–20.

———. *Imperial Gullies: Soil Erosion and Conservation in Lesotho.* Athens: Ohio University Press, 2005.

Souloumiac, Jean-Joseph. "La Déforestation au Maroc." CHEAM: unpublished report, 1948.

———. "Le Problème du défrichement ou une politique forestière au Maroc." CHEAM: unpublished report, 1947.

Spary, Emma C. *Utopia's Garden: French Natural History from Old Regime to Revolution.* Chicago: University of Chicago Press, 2000.

Sprugel, Douglas G. "Disturbance, Equilibrium, and Environmental Variability: What Is 'Natural' Vegetation in a Changing Environment?" *Biological Conservation* 58, no. 1 (1991): 1–18.

Steinheil, Ad. "Matériaux pour servir à la flore de Barbarie." *Annales des Sciences Naturelles* 9 (1838): 193–211.

Stoddart, David R. *On Geography and Its History.* Oxford: Basil Blackwell, 1986.

Surell, Alexandre. *Étude sur les torrents des hautes-Alpes, avec une suite par Ernest Cézanne.* 2nd ed. 2 vols. Paris: Dunod, 1870–72.

Swearingen, Will D. "Is Drought Increasing in Northwest Africa? A Historical Analysis." In *The North African Environment at Risk,* edited by Will D. Swearingen and Abdellatif Bencherifa, 17–34. Boulder, Colo.: Westview, 1996.

———. *Moroccan Mirages: Agrarian Dreams and Deceptions, 1912–1986.* Princeton, N.J.: Princeton University Press, 1987.

Swift, Jeremy. "Desertification: Narratives, Winners and Losers." In *The Lie of the Land: Challenging Received Wisdom on the African Environment,* edited by Melissa Leach and Robin Mearns, 73–90. London: International African Institute, 1996.

———. "Dynamic Ecological Systems and the Administration of Pastoral Development." In *Living with Uncertainty: New Directions for Pastoral Development in Africa,* edited by Ian Scoones, 153–73. London: Intermediate Technology Publications, 1995.

Tassy, Louis-François Victor. *Service forestier de l'Algérie, rapport adressé à m. le gouverneur de l'Algérie.* Paris: A. Hennuyer, 1872.

Terrier, Auguste. *Le Maroc.* Paris: Larousse, 1931.

Thinon, Michel, Aziz Ballouche, and Maurice Reille. "Holocene Vegetation of the Central Saharan Mountains: The End of a Myth." *Holocene* 6, no. 4 (1996): 457–62.

Thirgood, J. V. *Man and the Mediterranean Forest: A History of Resource Depletion.* New York: Academic Press, 1981.

Thomas, D. A., V. R. Squires, W. Buddee, and J. Turner. "Rangeland Regeneration in the Steppic Regions of the Mediterranean Basin." In *Rangelands: A Resource under Siege,* edited by P. L. Joss, 280–87. Cambridge: Cambridge University Press, 1986.

Thomas, David S. G. "Science and the Desertification Debate." *Journal of Arid Environments* 37, no. 4 (1997): 599–608.

Thomas, David S. G., and Nicholas Middleton. *Desertification: Exploding the Myth.* West Sussex: John Wiley, 1994.

Thomson, Ann. *Barbary and Enlightenment: European Attitudes towards the Maghreb in the 18th Century.* Leiden: E. J. Brill, 1987.

Tiffen, Mary, Michael Mortimore, and Francis Gichuki. *More People, Less Erosion: Environmental Recovery in Kenya.* Chichester: John Wiley, 1994.

Till, Claudine, and Joel Guiot. "Reconstruction of Precipitation in Morocco since 100 A.D. Based on *Cedrus atlantica* Tree-Ring Widths." *Quaternary Research* 33, no. 3 (1990): 337–51.

Trabaud, Louis. "Postfire Plant Community Dynamics in the Mediterranean Basin." In *The Role of Fire in Mediterranean-Type Ecosystems,* edited by José Moreno and Walter Oechel, 1–15. New York: Springer-Verlag, 1994.

Trabut, Louis. *L'Halfa.* Alger: Giralt, 1889.

———. "Les Régions botaniques et agricoles de l'Algérie." *La Revue Scientifique* 1, no. 15 (1881): 460–68.

Trautmann, Wolfgang. "The Nomads of Algeria under French Rule: A Study of Social and Economic Change." *Journal of Historical Geography* 15, no. 2 (1989): 126–38.

Trolard, Paulin. *Bréviaire du ligeur.* Alger: Casabianca, 1893.

———. *La Colonisation et la question forestière.* Alger: Casabianca, 1891.

———. "Recherches sur l'anatomie du système veineux de l'encéphale et du crâne." Thèse pour le Doctorat en Médecine, Faculté de Médecine de Paris, 1868.

Trottier, François. *Boisement dans le désert et colonisation.* Alger: F. Paysant, 1867.

———. *Boisement et colonisation: Rôle de l'eucalyptus en Algérie au point de vue des besoins locaux, de l'exportation, et du développement de la population.* Alger: Association Ouvrière, 1876.

———. *Notes sur l'eucalyptus et subsidiairement sur la nécessité du reboisement de l'Algérie.* Alger: F. Paysant, 1867.

Tsiouvaras, C. N., B. Noitsakis, and V. P. Papanastasis. "Clipping Intensity Improves Growth Rate of Kermes Oak Twigs." *Forest Ecology and Management* 15, no. 3 (1986): 229–37.

Turner, Matthew D. "Methodological Reflections on the Use of Remote Sensing and Geographic Information Science in Human Ecological Research." *Human Ecology* 31, no. 2 (2003): 255–79.

———. "Overstocking the Range: A Critical Analysis of the Environmental Science of Sahelian Pastoralism." *Economic Geography* 69, no. 4 (1993): 402–21.

United Kingdom. Naval Intelligence Division. *Algeria.* Geographical Handbook Series. Oxford: Oxford University Press, 1943.

———. *Morocco.* Geographical Handbook Series. Oxford: Oxford University Press, 1941.

———. *Tunisia.* Geographical Handbook Series. Oxford: Oxford University Press, 1945.

United Nations Educational, Scientific, and Cultural Organization. *Bioclimatic Map of the Mediterranean Zone. Explanatory Notes.* Vol. 21 of *Ecological Study of the Mediterranean Zone.* Paris: UNESCO–FAO, 1963.

———. *Vegetation Map of the Mediterranean Zone. Explanatory Notes.* Vol. 30 of *Ecological Study of the Mediterranean Zone.* Paris: UNESCO–FAO, 1969.

Valensi, Lucette. *On the Eve of Colonialism: North Africa before the French Conquest.* Translated by Kenneth J. Perkins. New York: Africana, 1977.

Vatin, Jean-Claude. "Désert construit et inventé, Sahara perdu ou retrouvé: Le Jeu des imaginaires." In *Le Maghreb dans l'imaginaire français: La Colonie, le désert, l'exil,* edited by Jean-Robert Henry, 107–31. Aix-en-Provence: La Calade, 1985.

Vaxelaire, Daniel, ed. *La Grande encyclopédie du Maroc.* Rabat: GEI, 1987.

Verne, Henri. *La France en Algérie.* Paris: Charles Douniol, 1869.

Voisin, Georges. *L'Algérie pour les Algériens.* Paris: Michel Lévy Frères, 1861.

Wachi, Commandant P. *La Question des forêts en Afrique.* Tunis: Imprimerie Rapide, 1899.

Wagstaff, J. Malcolm. *The Evolution of Middle Eastern Landscapes: An Outline to A.D. 1840.* London: Croom Helm, 1985.

Warnier, Auguste. *L'Algérie devant l'empereur, pour faire suite à l'Algérie devant le sénat, et à l'Algérie devant l'opinion publique.* Paris: Challamel Ainé, 1865.

———. *L'Algérie devant le sénat.* Paris: Dubuisson, 1863.

Warren, Andrew. "Changing Understandings of African Pastoralism and the Nature of Environmental Paradigms." *Transactions of the Institute of British Geographers* 20, no. 2 (1995): 193–203.

Webb, James L. A. *Desert Frontier: Ecological and Economic Change along the Western Sahel, 1600–1850.* Madison: University of Wisconsin Press, 1995.

———. *Tropical Pioneers: Human Agency and Ecological Change in the Highlands of Sri Lanka, 1800–1900.* Athens: Ohio University Press, 2002.

Weiner, Douglas R. "A Death-Defying Attempt to Articulate a Coherent Definition of Environmental History." *Environmental History* 10, no. 3 (2005): 404–20.

Wengler, Luc, and Jean-Louis Vernet. "Vegetation, Sedimentary Deposits and Climates during the Late Pleistocene and Holocene in Eastern Morocco." *Palaeogeography, Palaeoclimatology, Palaeoecology* 94, no. 1–4 (1992): 141–67.

Westoby, Mark, Brian Walker, and Immanuel Noy-Meir. "Opportunistic Management of Rangelands Not at Equilibrium." *Journal of Range Management* 42, no. 4 (1989): 266–74.

White, Frank. *The Vegetation of Africa: A Descriptive Memoir to Accompany the UNESCO/AETFAT/UNSO Vegetation Map of Africa.* Paris: UNESCO, 1983.

Whited, Tamara L. *Forests and Peasant Politics in Modern France.* New Haven, Conn.: Yale University Press, 2000.

Woolsey, Theodore. *French Forests and Forestry: Tunisia, Algeria, Corsica.* New York: John Wiley, 1917.

Zaim, Fouad. "Déforestation et érosion dans le Maroc méditérranéen, effets socio-économiques." In *La Forêt marocaine: Droit, économie, écologie,* edited by SOMADE, 95–109. Casablanca: Afrique Orient, 1989.

Zaimeche, Salah E. "The Consequences of Rapid Deforestation: A North African Example." *Ambio* 23, no. 2 (1994): 136–40.

Index

Page numbers in italics refer to figures. The letter *n* following a page number refers to a note on that page. It is followed by the note number. The abbreviation *pl.* refers to a color plate. It is followed by the plate number.